Moving Beyond Gridlock

Traffic and Development

No Longer Property of
EWU Libraries

No Longer Property of
EWU Libraries

Principal Author

Robert T. Dunphy
ULI–the Urban Land Institute

Contributing Authors

Deborah L. Brett
Deborah L. Brett & Associates

Sandra Rosenbloom
Drachman Institute/University of Arizona

Andre Bald
U.S. Peace Corps, Republic of the Philippines

Urban Land Institute

SCL
HE308
.D86
1997

About ULI–the Urban Land Institute

ULI–the Urban Land Institute is a nonprofit education and research institute that is supported and directed by its members. Its mission is to provide responsible leadership in the use of land in order to enhance the total environment.

ULI sponsors educational programs and forums to encourage an open international exchange of ideas and sharing of experience; initiates research that anticipates emerging land use trends and issues and proposes creative solutions based on this research; provides advisory services; and publishes a wide variety of materials to disseminate information on land use and development.

Established in 1936, the Institute today has some 13,000 members and associates from more than 50 countries representing the entire spectrum of the land use and development disciplines. They include developers, builders, property owners, investors, architects, public officials, planners, real estate brokers, appraisers, attorneys, engineers, financiers, academics, students, and librarians. ULI members contribute to higher standards of land use by sharing their knowledge and experience. The Institute has long been recognized as one of America's most respected and widely quoted sources of objective information on urban planning, growth, and development.

Richard M. Rosan
Executive Vice President

Project Staff

Senior Vice President, Research, Education, and Publications
Rachelle L. Levitt

Vice President/Publisher
Frank H. Spink, Jr.

Project Director
Robert T. Dunphy

Managing Editor
Nancy H. Stewart

Copy Editor
Libby Howland

Book Design/Layout
Helene Y. Redmond
HYR Graphics

Production Manager
Diann Stanley-Austin

Word Processor
Maria-Rose Cain

Recommended bibliographic listing:
Dunphy, Robert T., et al. *Moving Beyond Gridlock: Traffic and Development*. Washington, D.C.: ULI–the Urban Land Institute, 1997.

ULI Catalog Number: T03
International Standard Book Number: 0-87420-803-3
Library of Congress Catalog Card Number: 96-61394

Copyright ©1997 by ULI–the Urban Land Institute
1025 Thomas Jefferson Street, N.W.
Suite 500 West
Washington, D.C. 20007-5201

Printed in the United States of America. All rights reserved. No part of this book may be reproduced in any form or by any means, electronic or mechanical, including photocopying, recording, or by any information storage and retrieval system, without written permission of the publisher.

EASTERN WASHINGTON
UNIVERSITY LIBRARIES
CHENEY, WA 99004

Contents

Preface

What is the transportation problem? In most urban areas, it is enormously complex and not well understood by the experts, let alone by the public. Failure of public agencies to resolve critical transportation concerns can cause the problem to spill over into spheres ill-equipped to handle it, including the private sector. In many regions of the country that are not in compliance with federal air quality standards, controversial employee trip reduction programs have been advocated with the major burden for implementation assigned to private sector employers.

Having a reputation as an area with major transportation deficiencies can seriously hinder a community's, or a city's, or a region's ability to attract new businesses and residents. Traffic congestion degrades the quality of life for residents and is the leading reason for most grass-roots campaigns against specific development proposals and against growth in general. Gridlock on the roads can lead to gridlock in the planning office, delay and even scuttle development projects, and possibly embroil developers in legal battles.

This is not a how-to book. Recognizing the enormous challenge of delivering transportation services that meet private needs with largely public funds, it attempts to put the transportation problem in the context of regional land use and development decisions, and to offer some guidance to those responsible for making transportation improvements and to community planners and developers who approve and implement the projects that need transportation services.

There are probably no examples of large urban regions that have truly solved the transportation problem, whatever that means. However, the case study research in this book identifies some of the policies that make things worse, explores some of the practices that hold promise for improving the relationship between transportation and land use, and examines the experiences of some U.S. and Canadian urban areas that have accepted the challenge of delivering transportation.

The first section looks at what is the urban transportation problem from a number of angles—including its nature, its causes, and the responsibility for its solution. Chapter 1 describes the problem quantitatively. It reviews data on the causes of traffic growth; statistical comparisons of U.S. urban areas over 1 million people; and variations in supply, demand, and density measures for major modes of travel. Chapter 2 takes a look at how real estate projects and major transportation projects are developed, comparing the players, the process, the time lines, and the politics. Chapter 3 looks at major demographic trends and what they imply for travel into the next century.

The second section examines the practical experiences of a number of large metropolitan regions that have scored some remarkable successes in dealing with traffic and growth. The seven case studies are of regions that have taken a series of significant steps to relate transportation and land use policies more effectively. They include three areas—Portland, San Diego, and Toronto—that have emphasized regional development policies and transit. Phoenix is included for its urban villages growth management initiative in the early 1980s and its bold freeway development program in the second half of the decade. Two of the regions—Houston and Atlanta—have concentrated on increasing transportation investments and meeting private sector needs, with a lesser focus on regional growth policy. These six regions have all experienced high rates of growth in recent years. The seventh region—St. Louis—has not. However, despite its stagnant regional economy, St. Louis has growth centers, transportation investment needs, and a new and highly praised rail transit system, by which it hopes to catalyze activity in a reenergized core.

A wrap-up chapter summarizes some of the ideas from these (and other) experiences. ULI's hope is that this book will assist developers and transportation providers in dealing with future transportation needs.

Acknowledgments

The contributors to this book, the latest product of ULI's program of research on the relationship between land use and transportation, are many. Special thanks are owed to the members of the book's advisory committee:

- William R. Eager, TDA, Inc., Seattle, Washington;
- Christopher B. Leinberger, Robert Charles Lesser & Co., Santa Fe, New Mexico;
- H. Pike Oliver, Interra, Newport Beach, California;
- Richard B. Peiser, University of Southern California, Los Angeles, California; and
- Edwin D. Wetmore, City of Saginaw (retired), Saginaw, Michigan.

Chaired by Bill Eager, the members of this committee participated in regular meetings on the project, on-site workshops, and manuscript reviews.

The following are contributing authors:

- Deborah Brett, the author of Chapter 3, "Demographics, Changing Preferences, and Travel," is a consultant who writes on demographic and market issues for the real estate industry.
- Sandra Rosenbloom, the author of Chapter 6, "Phoenix," is a professor at the University of Arizona at Tucson, and director of the Drachman Institute. She is recognized internationally for research on the travel implications of demographic changes in society, particularly the growth of nontraditional families.
- Andre Bald, author of Chapter 5, "Atlanta," coauthor of Chapter 4, "Portland," and researcher in general for the book, was an intern at ULI, with previous experience at the Prince George's County Parking Authority in Maryland. He is currently a Peace Corps volunteer in the Philippines.

The U.S. Environmental Protection Agency's Office of Policy Planning and Evaluation provided financial support for a 1994 workshop in Portland, part of its "Market Acceptance of Development Concepts to Reduce Transportation Needs"project. The advice of William Schroeer and Paula Van Lare of that office is greatly appreciated.

A ULI panel sponsored by the Bi-State Development Agency and many local civic, business, and government cosponsors provided me the opportunity for an intensive week of on-site research in St. Louis.

A number of senior officials of the U.S. Department of Transportation offered valuable guidance and information. These include Sheldon Edner, Susan Liss, and Elaine Murakami of the Federal Highway Administration; Samuel Zimmerman, then director of the Federal Transit Administration's Office of Planning and now with FHWA's Office of Environment and Planning; and Richard Steinmann, director of the FTA's Office of Policy Development.

The database on urbanized areas was developed with the assistance of James McDonnell (deceased) of the Federal Highway Administration. His efforts to improve the quality of regional transportation data are gratefully remembered. John Hartman of the Transportation Association of Canada provided data on urbanized areas in Canada.

ULI sponsored one-day regional workshops in Portland and San Diego in 1994, at which over 70 participants offered their insights into complex issues of local transportation and development policy. The success of these workshops is owed to their chairs—John Russell, (Russell Development, Portland) and Sanford Goodkin (Sanford R. Goodkin and Associates, San Diego)—and Terry Lassar in Portland and Mike Stepner, Gene Geritz, and George Franck in

San Diego, as well as to all who participated, including John K. Stewart of San Francisco.

Each case study relied on many people who provided extensive background data, official reports, and expert opinion, and who reviewed manuscripts. In particular, I would like to acknowledge the help of the following individuals:

- **Portland.** G.B. Arrington, Henry Markus, Phil Whitmore (currently with Metro), and Shawn Ferguson of Tri-Met; Dennis Yee, Sonny Condor, and Mary Weber of Metro; and Sheryl Tweete of the Portland Development Commission and John Kenward, the commission's first executive director, who is now retired.
- **Atlanta.** James Carson of CARTER, a real estate company; Jeff Rader of the Atlanta Chamber of Commerce; Richard Courtney and Jim Bohn of the Atlanta Regional Commission; Truman Hartshorn of Georgia State University; John Edwards of the RBA Group; and Jerry Pachucki and Laura Gillig of MARTA.
- **Phoenix.** Gina Trimble of the Papago Park Center, Salt River Project; and Steve Tate of the Maricopa Association of Governments.
- **St. Louis.** Richard Ward of Development Strategies, Inc.; Les Sterman, Glenn Griffin, and Donna Humphreys of the East-West Gateway Coordinating Council; Sarah Smith of the Bi-State Development Agency; and James Farrell of the Regional Commerce and Growth Association.
- **Toronto.** Robert Pringle of Metro Toronto; Dale Martin of the Ontario Ministry of Housing and Urban Affairs; Paul Stagl of Opus Development; Richard Soberman of the University of Toronto; B.G. Hutchinson of the University of Waterloo (since retired); Edward Sajecki of the City of Burlington; Stuart Smith of Oxford Development; and Juri Pill of the Toronto Transit Commission.
- **San Diego.** George Franck of the San Diego Association of Governments; Michael Stepner of the City of San Diego; Donald Cerone of CalMat Properties; William Lieberman of the Metropolitan Transit Development Board; Andy Schlaefli of Urban Systems Associates; Janice Weinrick of the Centre City Development Corporation; and William Scott of the Douglas Allred Company.
- **Houston.** Roger Hord and John Walsh of the Greater Houston Partnership; and Jerry Bobo of the Houston-Galveston Area Council.

Many other people not mentioned here helped, and I appreciate their contributions too. Let me specifically acknowledge the fine work of four other ULI staff members: Ron Treschitta, who prepared most of the charts, Ronnie Van Alstyne, Joan Campbell, and Rick Davis.

Robert T. Dunphy
Senior Director, Transportation and Infrastructure, ULI

Chapter 1

Sorting Out the Problem

The only way to solve the traffic problems of the country is to pass a law that only paid-for cars are allowed to use the highways.

— Will Rogers

The transportation problem can be enormously complex. It is not well understood by the experts, let alone by the public. Lack of understanding, however, keeps no one from proposing remedies. Because travel is such a common experience, virtually everyone has a pet solution. Among traffic engineers, a favorite definition of a traffic professional is: anyone, regardless of age, gender, or experience.

The conflict between public perception and technical reality frequently pits the experts, who often are engineers, against citizens and elected officials. Non-experts often put forward a single program—build more roads, install a light-rail line, or stop development—as a simplistic solution. Different professions have their own biases in the solutions department. Engineers have been criticized for building new roads to serve demand, rather than taking steps to control demand. Economists prefer pricing solutions. Environmentalists may oppose new roads on the grounds that they will encourage growth.

In any community, a thorough grasp of the transportation problem requires not only in-depth knowledge of the various quantifiable components—demand, supply, and congestion—but also a sense of the public's awareness of the problem and its support for possible options. Actions that may be popular may not help solve the transportation problem, and actions that may help may not be popular. This is a common dilemma.

This chapter dissects some of the common elements and causes of the transportation problem in urban areas, in order to lay a foundation for some of the solutions discussed later. It also presents selected data for major U.S. urban areas. Outside investors and developers can use comparative regional information to get a quick read on the strengths and weaknesses in local transportation systems, an

important factor in relocation and investment decisions. Local citizens can use such data to begin to ask the right questions about their own regional transportation needs.

How Bad Is the Traffic?

A rational approach to transportation planning begins with measuring how well people and goods get around at present. Transportation planning tends to be oriented to future conditions, because new projects take so long to build. But it is difficult to get the public interested in long-term plans if no problem exists currently. The public generally responds best when there is a crisis.

Traffic congestion became that crisis in the 1980s, especially in those new suburban centers that blossomed with offices, shops, and venues for cultural and recreational activities, typically at the interchanges of freeways built in the 1960s or 1970s. Joel Garreau's term "edge city" caught on as a description of some of the new suburban downtowns, such as Perimeter Center near Atlanta, the Galleria/Post Oak area in Houston, and the Mission Valley development in San Diego.[1]

This suburban development, said Robert Cervero, was based on the white collarization of America's industry, which made firms footloose and free to relocate (or start up) in the suburbs. The urban fringes of fast-growing Sunbelt and West Coast metropolises became "powerful magnets for luring new startup industries as well as long-standing corporate giants seeking more attractive working environments."[2]

The traffic attracted to edge cities and the new fringe beehives of business and commerce quickly expanded to fill up the remaining highway capacity, and caused lengthening traffic delays. Traffic congestion became a media topic, locally and nationally. Surveys in Houston, San Francisco, and many other communities showed that citizens felt traffic congestion was the worst problem facing their regions. Transportation professionals took up the issue of suburban traffic congestion. The Institute of Transportation Engineers convened a number of national conferences on the topic beginning in 1985. The experts encountered a disturbing problem: traffic engineers could calculate congestion at intersections and other bottlenecks, but there was no generally recognized measure of traffic congestion.

The term "congestion" is most useful in describing bottlenecks at individual intersections and major roadway merger points. When the number of vehicles seeking to go past a point exceeds the capacity, traffic backs up.[3] At an intersection controlled by a signal, this means that some vehicles will not get through the signal on the first green light. Traffic may back up from the intersection, possibly far enough to block the preceding intersection. In a grid street system such as a downtown, vehicles backed up into an intersection can block cross traffic, producing gridlock, a traffic nightmare in which nothing moves in any direction. On a freeway, if more vehicles seek to go past a point than it can handle, traffic backs up in a stop-and-go pattern.

Such congestion is tolerable during brief peak periods. In most cases, congestion occurs on only parts of the entire transportation system, so most trips do not involve constant bumper-to-bumper travel. From a regional perspective, eliminating all peak period congestion would be inappropriately expensive and self-defeating in that it would encourage people to travel during peak travel times. In fact, congestion serves as a means of regulating travel behavior. Controlled congestion has been called the key to a civilized society. Regions need to find the level of congestion that balances personal inconvenience and public costs, along with other criteria of regional transportation systems, such as safety and environmental protection.

The need for consistent measures of congestion, especially for some growing communities in Texas, led the College Station–based Texas Transportation Institute to develop in 1988 a method of aggregating congestion at various points into a regional congestion index. Using this method, Figure 1-1 documents the growth in congestion during the 1980s for most U.S. metropolitan areas. In 1991, the most congested city was, not surprisingly, Los Angeles, followed by Washington, D.C., San Francisco, Chicago, and Miami. In 1982, which was the first year of the survey, Houston and Phoenix were at the top of the list of most congested cities.

While this index of worsening congestion confirmed many people's fears, the contrast of reported congestion with people's own travel experiences was striking. The 1990 Census reported that the duration of the average U.S. commute rose modestly from slightly less than 22 minutes in 1980 to slightly more than 22 minutes.[4] On balance, the suburbanization of travel had actually raised the average speed of commuting trips between 1983 and 1990, as commuters switched from buses to cars and from city roads to suburban roads that carried them at higher speeds.[5]

Did worsening congestion in the 1980s affect development patterns? Many analysts expected that one consequence would be that people would move closer to work. Wrong. Surveys of households buying new homes during the 1980s revealed an average commuting distance of 16 to 17 miles, which is almost twice as long as the average work trip reported for all commuters in 1990. The average commuting time for these homebuyers was not much higher than the average time—22 minutes—for all commuters: 25 to 27 minutes for buyers of detached homes and 30 minutes for buyers of attached homes.[6] In a 1995 survey, potential buyers of new homes (most of whom were homeowners) were asked about the compromises they would be willing to make to be able to afford the house they wanted. Thirty-two percent were willing to settle for a longer-distance commute, and 34 percent were willing to live farther from shopping and entertainment. Only 24 percent would compromise on a smaller house.[7]

In the public's mind, congestion is associated with the commute. However, the importance of commuting in daily travel patterns is clearly on the decline. By 1990, the number of trips made for the purpose of earning a living, a travel category that includes commuting and other work-related travel, represented less than one-third of travel during the weekday evening peak period (4 p.m. to 7 p.m.).[8] Evening peak hour trips made for the purposes of family and personal business outnumbered work trips. The commute—making commuting by car during peak hours less attractive or making alternatives more attractive—has been at the center of transportation policy debates, while virtually no attention has been given to the nonwork trip. What can be done about the parent bringing kids to soccer practice, the worker heading for a health club, or the student attending evening classes? Providing practical, affordable alternatives to the private auto for such trips is a much more difficult problem, and it generally is not being addressed.

All in all, the buildup in traffic congestion during the 1980s was difficult to measure and its causes and cures were multiple. While the experts were prophesying a transportation crisis, travelers found

Figure 1-1

Change in Roadway Congestion, 1982–1991, and 1991 Congestion Values, Major U.S. Metropolitan Areas[1]

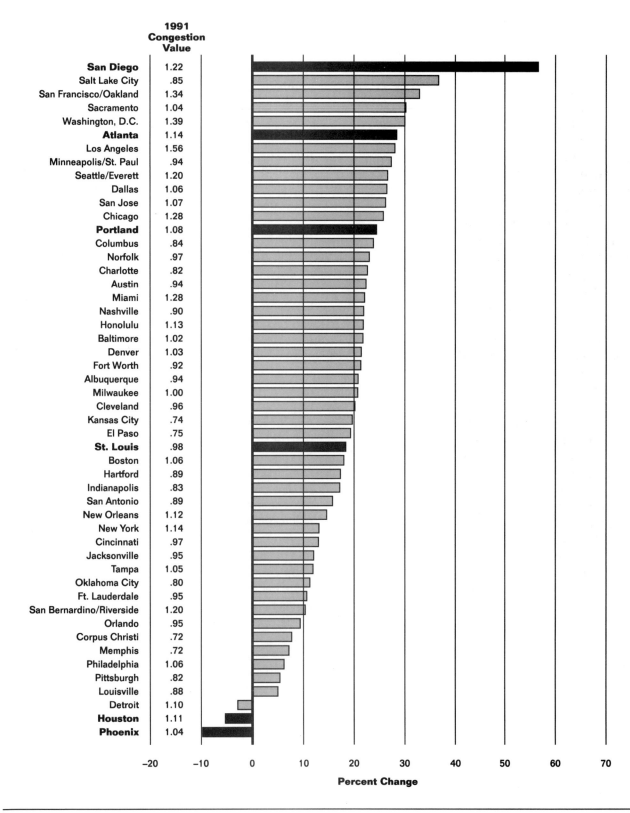

	1991 Congestion Value
San Diego	1.22
Salt Lake City	.85
San Francisco/Oakland	1.34
Sacramento	1.04
Washington, D.C.	1.39
Atlanta	1.14
Los Angeles	1.56
Minneapolis/St. Paul	.94
Seattle/Everett	1.20
Dallas	1.06
San Jose	1.07
Chicago	1.28
Portland	1.08
Columbus	.84
Norfolk	.97
Charlotte	.82
Austin	.94
Miami	1.28
Nashville	.90
Honolulu	1.13
Baltimore	1.02
Denver	1.03
Fort Worth	.92
Albuquerque	.94
Milwaukee	1.00
Cleveland	.96
Kansas City	.74
El Paso	.75
St. Louis	.98
Boston	1.06
Hartford	.89
Indianapolis	.83
San Antonio	.89
New Orleans	1.12
New York	1.14
Cincinnati	.97
Jacksonville	.95
Tampa	1.05
Oklahoma City	.80
Ft. Lauderdale	.95
San Bernardino/Riverside	1.20
Orlando	.95
Corpus Christi	.72
Memphis	.72
Philadelphia	1.06
Pittsburgh	.82
Louisville	.88
Detroit	1.10
Houston	1.11
Phoenix	1.04

Percent Change

[1]The TTI roadway congestion index is calculated as a ratio of freeway arterial VMT to standard arterial freeway capacities.
Source: Trends in Urban Roadway Congestion—1982 to 1991 (Texas Transportation Institute).

sufficient flexibility in the system to adjust their travel behavior—a solution that further demonstrates the complexity of the transportation problem, both political and technical.

The Causes of Traffic Growth In the 1980s

Growth in travel demand far outdistanced underlying population growth during the 1980s. While the U.S. population gained by a modest 4 percent between 1983 and 1990, the number of vehicle-miles traveled (VMT) jumped by 41 percent. Some powerful national trends explain the disproportionate growth in VMT: changing demographics, growing dependency on the automobile, and lengthening travel distances. Each of these components of driving demand accounted for approximately one-third of the large increase in household VMT. In short, more people made more and longer driving trips.

Changing Demographics

Overall population growth and a disproportionate increase in the number of people of peak driving age accounted for 36 percent of the travel increase. Not only were there more people, but people made more daily trips. The number of workers grew almost 250 percent faster than the population, as women's rate of labor force participation increased. The baby boom generation reached the age of peak travel. The rates of growth of vehicle ownership and of the work force were almost identical, suggesting that increased car dependency could be explained entirely by job growth. (Demographic influences on travel are addressed in more detail in Chapter 3.)

Growing Dependency on The Automobile

Greater reliance on cars for daily trips accounted for 25 percent of the growth in travel. During the 1980s, the family car—a single car shared by different drivers —virtually disappeared, and by 1990 the number of vehicles owned by households had surpassed the number of drivers within households. While the number of trips in private vehicle made by people as driver or passenger increased by 15 percent, the number of trips made by people as driver increased by 25 percent.

About half of this factor of increasing dependency on cars was due to a lower use of transit and half was because of less carpooling. Trips made by public transit declined by more than 10 percent from 1983 to 1990, when there were just below 5 billion trips, the same number as in 1969. (Between 1977 and 1983, transit ridership had increased.) Trips in school buses

declined slightly, probably reflecting fewer students and an increasing tendency of suburban children to drive or be driven to school.

As to the decline in shared driving during this period, some falloff could have been expected from the decline in family size: a smaller family means fewer people in the car, even when the entire family travels. In fact, shopping trips became slightly smaller (1.7 people in 1990, down from 1.8 in 1983), but the number of persons per other family-oriented trip remained constant (at 1.8 persons per family business trip and 2.1 persons per social or recreational trip). Smaller commuting trips made the biggest impact. The home-to-work trip, which had consistently averaged 1.3 persons per car as late as 1983, declined to only 1.1 persons by 1990. The number of commuting vehicles per household rose 8.2 percent.

Lengthening Travel Distances

Longer drives accounted for 38 percent of the growth in VMT. Travel distances increased despite the widespread suburbanization of jobs, services, and entertainment during the 1980s. Commuting distances lengthened by more than 25 percent—the largest increase. Shopping trips, bucking the trend, became slightly shorter, apparently reflecting the ability of retailers to follow the market. Longer travel distances and the increased dependency on cars were, in many respects, closely related to changing demographics and lifestyles. Baby boomers bought their homes in distant suburbs where there were no alternatives to driving. Many families were under such time pressure that even where alternative travel options might have been available they were just not practical.

The national data on the causes of travel demand suggest how difficult it will be to reduce driving by limiting growth, providing alternatives to driving, and planning communities to reduce the need to drive. People are not driving capriciously. They are driving more because they are more likely to be working (especially if they are women) or engaging in essential family travel. Efforts to reduce traffic growth by limiting housing development address only a small component of the problem. Housing growth limits would have to be combined with more aggressive policies to encourage travel by other modes and to effectively shorten the distances between destinations.

Declining Investment in Transportation

Much of the debate over traffic congestion focuses on demand issues—driving and development. Supply-side reasons for congestion are less appreciated. Much of the suburban development that took place in the 1980s was in corridors served by the U.S. interstate highway system and other major freeways completed in the 1970s. But no major expansion of highways took place to serve the new growth. Public funding helped rebuild the nation's transit systems in the 1980s, but most of these are oriented primarily to major downtowns, which have captured only a small share of recent travel growth.

Total spending on transportation by consumers, businesses, and governments provides the context for investment in transportation infrastructure. This U.S. transportation bill reached almost $1 trillion in 1992, or $3,970 per person, of which $2,500 was for passenger travel and $1,470 was for freight. Expenditures by federal, state, and local governments on transportation (land, sea, and air) maintenance, operations, and improvements amounted to about one-tenth of the total transportation bill.[9] The demand for transportation is largely private, but government is largely responsible for providing the underlying facilities. This inherent conflict brings to mind Will Rogers's solution to the traffic problem: "Let the government build the cars and the private sector build the roads."

From 1983 to 1993, annual traffic growth of 3.4 percent has continued to outstrip annual highway capacity, and road traffic has increased 47 percent on the average interstate highway and 30 percent on other main arterials in urban areas, compared with capacity increases of 30 percent and 2 percent, respectively. The severity of peak-hour congestion in U.S. urban areas has considerably worsened, especially in the 33 urban areas with over 1 million people. These large areas account for fully two-thirds of urban congestion, even though they account for only half of urban travel. Sixty percent of the mileage of principal arterial roads requires rehabilitation in the near future, and roads in large urban areas require some of the most expensive improvements.[10]

The slowdown in highway investment is a result of lack of funding, citizen opposition to highway projects, and new environmental and other regulations that have added to the complexity of project development. Gas taxes, the primary source of revenues for state and federal projects, are a problematic funding device. Sharp declines in the average fuel consumption of personal vehicles, in part due to federal policies designed to promote energy independence, have reduced revenues based on a fixed tax per gallon, which is the structure of most federal and state gas taxes. (In the 1980s, some states, thinking that the run-up in gasoline prices caused by OPEC supply constraints would repeat itself, pegged the gas tax to price rather than volume, and watched in dismay as prices went through the floor.) While revenues earmarked

for transportation improvements have not kept pace, the costs of road construction and maintenance have continued to inflate.

Urban transit's situation has improved, by national standards. Transit systems, which were once mostly private services, have been taken over by and large by the public sector. Transit facilities have improved markedly since the mid-1980s. The average age of equipment has stayed the same, operating speed has increased 10 percent, and the condition of maintenance facilities, stations, and bridges has improved greatly. The number of rail maintenance yards rated in "good" or "better" condition rose from 17 percent in 1984 to 64 percent in 1992.[11]

Recent investment in transit infrastructure seems to have been remarkably successful in modernizing facilities and improving services. While bus ridership is declining, the trend for rail ridership is up, in part due to the opening of some new systems. Federal and state funding has largely stabilized the capital stock, but most rail systems are heavily subsidized and their place in the regional transportation picture is being reassessed. The travel markets that rail systems serve best—downtown workers and central city residents and visitors—are in decline. Conventional transit cannot easily serve the fastest-growing market, which is people traveling from one suburb to another.

Public funding of transportation facilities waxes and wanes according to prevailing attitudes toward government spending. At times, the electorate prefers spending on social programs; at other times, it prefers spending on economic development and infrastructure programs; and sometimes it prefers no spending. Public works spending as a share of total spending at all levels of government declined from 20 percent in 1950 to 7 percent in 1984, while spending on welfare and education increased from 10 percent to over 40 percent of total expenditures.[12] Most government spending on transportation is supported by revenues from users, but elected officials are still reluctant to increase user fees because fee hikes are considered to be tax increases. Reluctance to raise user fees is a significant problem, and the potential shift from gasoline to alternate fuels may jeopardize future funding.

Limited funding is not the only supply-side constraint. Transportation advocates face a growing challenge in the form of popular opposition to new or expanded transportation facilities. While highway projects are most often objected to, proposals for new rail lines are hardly immune to opposition. New residents of an area sometimes object to the intrusion of the roads planned to serve them, and in some cases long-planned facilities have been removed from the maps. Such inconsistencies in planning are a fact of

Figure 1-2

Transportation Indicators for Major U.S. and Canadian Urbanized Areas

	1990 Population (Thousands)	Area (Square Miles)	Density Persons per Square Mile	Density Persons per Acre	All Roadways Total Miles	All Roadways Miles per 1,000 Persons	Freeways Miles Total	Freeways Miles As a Percent of All Roadways	Freeways Lane-Miles Total	Freeways Lane-Miles Per 1,000 Persons	Arterials (Miles)
New York/Northeast New Jersey	15,780	3,177	4,967	7.76	35,276	2.2	1,039	2.9%	5,747	0.36	6,125
Los Angeles/Long Beach	11,428	2,100	5,442	8.50	25,073	2.2	595	2.4	4,722	0.41	5,682
Chicago/Northwest Indiana	7,303	1,958	3,730	5.83	19,309	2.6	410	2.1	2,419	0.33	3,157
Philadelphia	4,216	1,086	3,882	6.07	10,805	2.6	297	2.7	1,453	0.34	2,156
Detroit	3,905	1,243	3,142	4.91	12,605	3.2	281	2.2	1,772	0.45	2,313
San Francisco/Oakland	3,676	816	4,505	7.04	9,008	2.5	343	3.8	2,143	0.58	1,604
Toronto	3,667	572	6,409	10.01	n/a	n/a	n/a	n/a	1,243	0.34	n/a
Washington, D.C.	3,240	820	3,951	6.17	8,556	2.6	287	3.4	1,669	0.52	1,401
Dallas/Fort Worth	3,030	1,404	2,158	3.37	18,946	6.3	436	2.3	2,536	0.84	1,699
Boston	2,803	1,033	2,713	4.24	9,323	3.3	257	2.8	1,471	0.52	1,783
Houston	2,798	1,549	1,806	2.82	17,001	6.1	313	1.8	1,945	0.70	1,441
Montreal	2,437	373	6,533	10.21	n/a	n/a	n/a	n/a	1,128	.046	n/a
San Diego	2,294	680	3,374	5.27	5,698	2.5	230	4.0	1,655	0.72	1,007
Minneapolis/St. Paul	2,055	996	2,063	3.22	8,951	4.4	290	3.2	1,452	0.71	1,048
Baltimore	1,991	523	3,807	5.95	5,948	3.0	235	4.0	1,233	0.62	918
St. Louis	1,950	694	2,810	4.39	7,164	3.7	269	3.8	1,551	0.80	1,208
Phoenix	1,920	971	1,977	3.09	9,396	4.9	85	0.9	584	0.30	1,267
Atlanta	1,860	1,198	1,553	2.43	9,691	5.2	264	2.7	1,875	1.01	1,226
Miami	1,800	442	4,072	6.36	5,602	3.1	107	1.9	656	0.36	528
Cleveland	1,752	629	2,785	4.35	5,536	3.2	223	4.0	1,181	0.67	904
Seattle/Everett	1,730	645	2,682	4.19	6,472	3.7	174	2.7	965	0.56	993
Pittsburgh	1,708	713	2,396	3.74	7,565	4.4	215	2.8	860	0.50	1,460
Vancouver	1,632	1,017	1,605	2.51	n/a	n/a	n/a	n/a	652	0.40	n/a
Denver	1,540	433	3,557	5.56	5,930	3.9	165	2.8	879	0.57	922
San Jose	1,410	326	4,325	6.76	3,714	2.6	166	4.5	1,000	0.71	653
Kansas City	1,281	608	2,107	3.29	6,207	4.8	304	4.9	1,497	1.17	1,003
Milwaukee	1,220	550	2,218	3.47	4,788	3.9	106	2.2	571	0.47	1,089
Ft. Lauderdale/Hollywood	1,200	368	3,261	5.10	4,208	3.5	85	2.0	588	0.49	402
Portland	1,196	416	2,875	4.49	4,514	3.8	127	2.8	625	0.52	711
San Bernardino/Riverside	1,169	480	2,435	3.81	3,750	3.2	121	3.2	738	0.63	711
San Antonio	1,165	442	2,636	4.12	6,730	5.8	152	2.3	833	0.72	614
Cincinnati	1,136	564	2,014	3.15	3,787	3.3	158	4.2	877	0.77	543
Sacramento	1,096	340	3,224	5.04	3,495	3.2	97	2.8	641	0.58	482
New Orleans	1,080	361	2,992	4.67	2,982	2.8	61	2.0	330	0.31	468
Buffalo	1,064	405	2,627	4.10	3,585	3.4	145	4.0	644	0.61	676
St. Petersburg	1,020	554	1,841	2.88	4,241	4.2	43	1.0	185	0.18	263
Average for Selected Cities	2,793	847	3,180	4.97	8,218	3.3	224	2.7	1,398	0.56	1,219

life for public officials, and they are not the only kind of policy inconsistency that creates traffic problems.

Other Contributing Factors

The challenge of providing transportation facilities to serve growth and changing lifestyles and travel needs is made even more difficult by a variety of private actions and official policies that exacerbate traffic congestion by encouraging driving or discour-

aging alternatives to driving. Among these are the following common practices.

Zoning to Exclude Moderate-Income Housing

In certain upscale communities, elitist zoning keeps out the type of housing local service workers can afford, forcing them to travel longer distances to work. Excessive minimum lot sizes can force people with sufficient income to buy a larger house or yard than

Figure 1-2

Transportation Indicators for Major U.S. and Canadian Urbanized Areas (continued)

	Roads	Driving						
		Daily Vehicle-Miles Traveled		Daily VMT[1] on Freeways			Transit Use	
	Freeway and Arterial Miles (Per 1,000 Persons)	Total (Thousands)	Per Capita	VMT (Thousands)	Percent of Total VMT	Per Freeway Lane-Mile	Total Trips 1991 (Millions)	Trips per Capita 1990
New York/Northeast New Jersey	0.4	225,010	14.3	80,917	36%	14,225	2807.6	175.0
Los Angeles/Long Beach	0.5	250,673	21.9	106,678	43	22,823	515.4	45.2
Chicago/Northwest Indiana	0.4	123,471	16.9	34,438	28	14,947	697.2	102.6
Philadelphia	0.5	65,759	15.6	18,278	28	12,510	367.5	87.0
Detroit	0.6	78,223	20.0	26,683	34	15,075	94.9	25.7
San Francisco/Oakland	0.4	76,950	20.9	41,972	55	19,191	414.1	114.1
Toronto	0.0	20,629	11.0	n/a	n/a	n/a	496.8	135.5
Washington, D.C.	0.4	64,323	19.9	25,020	39	15,321	372.4	110.7
Dallas/Fort Worth	0.6	80,200	26.5	33,462	42	13,185	56.6	17.7
Boston	0.6	51,338	18,3	22,073	43	15,046	323.7	116.6
Houston	0.5	71,613	25.6	28,294	40	14,751	90.7	31.2
Montreal	n/a	n/a	n/a	n/a			385.8	158.3
San Diego	0.4	51,606	22.5	26,758	52	16,405	68.6	29.2
Minneapolis/St. Paul	0.5	43,185	21.0	16,862	39	11,766	69.6	33.5
Baltimore	0.5	36,374	18.3	15,182	42	12,423	116.7	61.7
St. Louis	0.6	45,288	23.2	18,717	41	12,233	44.6	22.9
Phoenix	0.7	39,654	20.7	7,052	18	14,104	32.1	16.0
Atlanta	0.7	64,831	34.9	24,594	38	13,286	149.6	69.3
Miami	0.3	33,526	18.6	10,189	30	15,748	77.8	40.6
Cleveland	0.5	32,289	18.4	13,210	41	11,348	74.7	44.5
Seattle/Everett	0.6	40,840	23.6	16,357	40	16,862	100.3	57.5
Pittsburgh	0.9	32,466	19.0	7,180	22	7,933	89.3	53.2
Vancouver	0.0	n/a	n/a	n/a	n/a	n/a	104.2	63.9
Denver	0.6	27,150	17.6	10,733	40	12,869	55.5	36.6
San Jose	0.5	32,445	23.0	15,540	48	16,409	45.7	31.9
Kansas City	0.8	27,468	21.4	12,373	45	9,044	18.8	14.8
Milwaukee	0.9	28,659	23.5	7,513	26	13,517	66.3	54.1
Ft. Lauderdale/Hollywood	0.3	24,299	20.2	5,716	24	11,120	18.6	15.0
Portland	0.6	22,416	18.7	8,623	38	13,998	58.9	50.3
San Bernardino/Riverside	0.6	25,049	21.4	11,927	48	16,360	9.8	8.4
San Antonio	0.5	25,317	21.7	8,477	33	10,689	42.1	37.3
Cincinnati	0.5	23,612	20.8	10,893	46	12,549	34.9	28.7
Sacramento	0.4	23,619	21.6	8,848	37	13,977	20.3	18.5
New Orleans	0.4	16,723	15.5	3,886	23	11,704	85.5	82.2
Buffalo	0.6	17,003	16.0	5,369	32	8,590	30.4	31.8
St. Petersburg	0.3	17,982	17.6	1,399	8	7,441	19.8	11.6
Average for Selected Cities	0.5	51,111	19.2	19,034	33	12,429	223.8	56.5

[1] Vehicle-miles traveled.

Sources: 1990 Highway Statistics (Federal Highway Administration); *1991 Transit Profiles* (Federal Transit Adminstration); and *Urban Transportation Indicators in Eight Canadian Urban Areas* (Transportation Association of Canada, 1996).

they want, requiring more land to accommodate population growth, causing the region to expand in area, and creating longer travel distances.

Zoning to Exclude Housing

Many communities engage in fiscal zoning. They compete for office, business park, and retail development that will bring in jobs and tax revenues, while they discourage residential development because it brings in school-age children who cost more to edu-

cate than their families pay in taxes. As a result, the people who have jobs in these communities must commute from longer distances.

Pedestrian Barriers

People in certain locations might consider leaving the car home for some travel purposes, if their walking or bicycling access to destinations were not blocked by barriers of one kind or another. Walled communities, which are popular in southern California and

Texas, constitute such a barrier. In some cases, impassable streams, fences, or walls cut off commercial areas from nearby neighborhoods. Circuitous streets and many culs-de-sac in neighborhoods can make routes to nearby destinations too indirect; some residents may live close enough to see the nearest shops, but too far to walk. Some site designs virtually guarantee that the only access is by car. The classic suburban style commercial development—a box in an island of parking—is usually set back too far and often too dangerous for all but the most adventurous to access on foot.

Parking Subsidies

Subsidized parking can make driving cheaper than transit, especially for downtown workers. Many employers in downtown Los Angeles, Washington, D.C., and even New York City offer parking. In suburban workplaces, parking is virtually universal and is, in fact, one of the major attractions of a suburban job.

Uniform Pricing

Peak hour travel is not discouraged by the uniform pricing of street travel. On most roads, travel is free, which is to say that the driver pays no charges beyond the (not inconsiderable) cost of owning and operating the vehicle. According to a study by the Bay Area Economic Forum, the cost to all road users of adding a car to the traffic stream can vary tremendously depending on the time and place, from $0.60 per mile for peak hour drivers in the core area to mere pennies for reverse commuters.[13]

Land use and public policy practices that complicate the traffic problem are not for that reason any easier to change. They all have their reasons. Anthony Downs points out that such practices collectively serve "the dominant vision of unlimited low-density sprawl," which is actually the American ideal. They also are the source of many of our mobility problems, he adds, in that they prevent the formulation of effective regional policies to deal with traffic and also isolate low-income and minority families in inner-city ghettos.[14]

Regional Comparisons

The determination of what transportation services are appropriate for a particular metropolitan region depends or should depend on where the region is heading and what its regional vision is. In the view of many people, the transportation problem is too important to be left to transportation experts. No general statement of the transportation problem and no single set of possible solutions will fit all regions. It is necessary to deal with the specifics of each region.

Just as all politics is local, all transportation and growth problems are local. However, comparisons that show how a particular region stacks up against other regions in terms of various land use, growth, and transportation indicators can offer some insight, perspective, and possibly practical ideas, models, and paths to avoid.

Figure 1-2 pulls together a variety of transportation indicators for 34 U.S. and Canadian urban areas with more than 1 million people. (The author used the comparative data to help select the cities for case studies for this book.) Why only large regions? Business executives, developers, and investors are most interested in these regions. Also, the land use and transportation issues tend to be more complex, usually involving a greater number of governments and more difficult interagency conflicts. Smaller regions often look to larger communities for ideas. In large regions, the range of realistic transportation options is much greater. Finally, current data is generally more available. (Individuals interested in making some of the same comparisons for smaller regions can obtain data from the sources listed in Figure 1-2.)

Roads and Driving

The most basic measure of the level of road access is miles of roads. In urban areas, most roads are local streets. Among the 62 urbanized areas with over 200,000 people in 1993, local roads serving primarily as access to property accounted for 71 percent of total miles of roads. Principal highways and arterials accounted for only 9 percent of the miles of roads, but they carried 61 percent of the daily travel.[15] Regions with the largest number of miles of principal roads (freeways and arterial streets) on a per capita basis in 1990 were Pittsburgh, Milwaukee, Kansas City, Phoenix, and Atlanta. At the low end were three urbanized areas in Florida—Miami, Tampa/St. Petersburg, and Fort Lauderdale/Hollywood. Also near the bottom of the list were two urbanized areas in California—San Francisco and San Diego—along with New York and Chicago.

A region's freeway orientation can be measured in two ways: the share of traffic carried on freeways and the number of miles of freeway lanes per person. When ranked by the share of highway travel on freeways, an indicator of demand, four California regions topped the list. In San Francisco and San Diego, slightly more than half of all highway travel was on freeways in 1990, and freeway orientation was almost as heavy in San Jose and San Bernardino/Riverside. Outside California, the urban areas with the heaviest orientation of travel to freeways were Cincinnati, Kansas

City, and, surprisingly, Boston. The areas that were least dependent on freeways in 1990 were Tampa/ St. Petersburg, where freeways carried only 8 percent of traffic, and Phoenix, where freeways accounted for 18 percent of travel. (In most areas, these ratios change slowly, but Phoenix's aggressive freeway building program, described in Chapter 6, helped increase that region's share of daily travel on freeways to 22 percent by 1994.) Pittsburgh, Fort Lauderdale, and New Orleans were also on the low end of this ranking of major urbanized areas by the share of traffic carried on freeways. In most regions, 30 to 40 percent of travel was on freeways.

The rankings of these regions by the size of their freeway systems as measured in freeway lane-miles per capita were different. By this measure, an eight-lane section counts twice as much as a four-lane section. None of the top five regions was in California. Three of the top five were in the Midwest—Kansas City, St. Louis, and Cincinnati—and the other two were Atlanta (the second highest) and Dallas/Fort Worth. San Diego ranked sixth, the highest ranking California area. The urbanized areas with the fewest freeway lane-miles were Phoenix, Tampa/St. Petersburg, New Orleans, Chicago, Philadelphia, New York, Miami, and Los Angeles (contrary to popular perception). Toronto's freeway capacity in 1991 was estimated as equal to that of the Philadelphia area.

The average number of miles per capita driven daily—the driving habit—is a measure of auto dependency. As shown in Figure 1-3, in 1990 the driving habit ranged from 35 miles daily in Atlanta to 14 miles per day per resident in New York. Two urban areas in Texas—Dallas and Houston—are unsurprisingly among the top five cities in this category, along with Seattle and Milwaukee, with St. Louis ranking a close sixth. Older, high-density regions like New Orleans, Philadelphia, Buffalo, and Chicago tend to be low on this list. The spread between top-ranked and bottom-ranked cities is not as great in this category as in some others, such as miles of freeway lanes or transit use. Residents of Milwaukee, for example, the fifth highest urban area in terms of per capita driving, averaged only about one-third more daily miles than the residents of Chicago, the fifth lowest urban area.

Transit

U.S. metropolitan regions have extreme variations in transit service and use. The New York region's extensive network of subways, commuter trains, and buses carries almost 3 million riders every weekday. In 1990, the New York metropolitan area accounted for 37 percent of all transit commuters in the United States.[16] Regional transit comparisons begin with the observation that there is New York and then there is not exactly New York. Toronto's transit system plays a similar role in Canada, although several other urban areas in Canada have high levels of transit use.

Six U.S. regions other than New York had more than 250,000 daily transit riders in 1990, and their combined daily ridership almost equals that of New York. The regions with the highest volume of transit ridership were the most populous regions—except for Detroit and Dallas/Fort Worth, both of which had fewer than 250,000 daily riders.

On a per capita basis, New York's transit habit was also impressive, with 175 annual trips per resident as shown in Figure 1-4, ahead of Montreal and Toronto. Boston, San Francisco/Oakland, Washington, D.C., and Chicago followed, with 103 to 117 annual transit trips per capita in 1990. Transit use was also fairly heavy in Philadelphia, New Orleans, and Atlanta, where annual rides per capita ranged between 69 and 87. While New York had the strongest transit habit and the weakest driving habit, Atlanta was near the top of both lists, demonstrating that people who drive more do not necessarily use transit less and vice versa. In 1990, the five urbanized areas of over 1 million population with the lowest levels of transit use were: Tampa/St. Petersburg, San Bernardino/Riverside, Kansas City, Fort Lauderdale/Hollywood, and Phoenix.

Density

Density—the number of people per unit of land—is a measure of urban form. A high-density pattern of development tends to be associated with shorter driving distances and more potential for efficient transit facilities. Any large region is composed of dense urban settlements, open land, and a mixture of densities in between. The relative density of a city can be measured by the eye, but it is not as easy to see the density of an entire region.

In a 1990 ranking of urban areas by overall density, New York was near the top but actually outranked by Los Angeles—despite southern California's reputation as the model of a spread-out lifestyle. Two other California regions, San Francisco and San Jose, ranked next, ahead of such East Coast cities as Miami, Washington, D.C., Philadelphia, Baltimore, and Chicago. The urbanized areas on the top of the density list were more than twice as densely populated as those on the bottom, which included four Sunbelt cities—Atlanta, Houston, Tampa/St. Petersburg, and Phoenix—and two Midwest cities—Cincinnati and Minneapolis/St. Paul.

Washington, D.C., and San Francisco had urbanized area populations of about the same size and both regions occupied about 800 square miles of land. In

Figure 1-3

Vehicle-Miles Traveled per Capita, 1990, Major U.S. Urbanized Areas

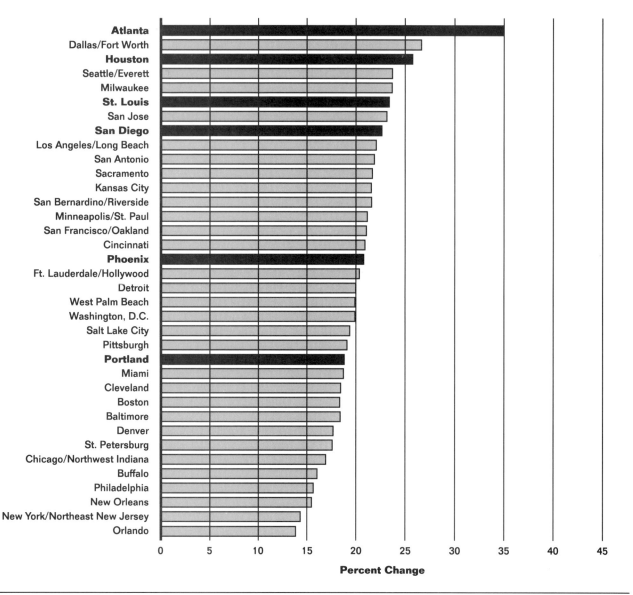

Source: 1990 Highway Statistics (Federal Highway Administration).

contrast, metropolitan Atlanta had half the population but its urbanized area was 50 percent larger, and Houston, with a slightly smaller population, spread out over twice as much land.

Cross-Comparisons

The real value of a series of who's-on-first rankings is that it enables people to make cross-comparisons that interrelate different factors. Not only transportation experts but also business and community leaders and others can develop insights on the relationships between transportation service, growth, and traffic. For example, an examination of the data

shows that the most congested cities are not necessarily the cities with the biggest driving habit or with the fewest roads:

- Los Angeles may have one of the most extensive freeway systems in the United States, but the region ranked among the lowest third in number of miles of freeway lanes per capita. The region's driving index was only slightly above average, but Los Angeles topped the congestion chart.
- Atlanta's number one ranking in per capita miles driven was matched by its extensive freeway system (second from the top), and thus the

Figure 1-4
Transit Trips per Capita, 1990, Major U.S. Urbanized Areas

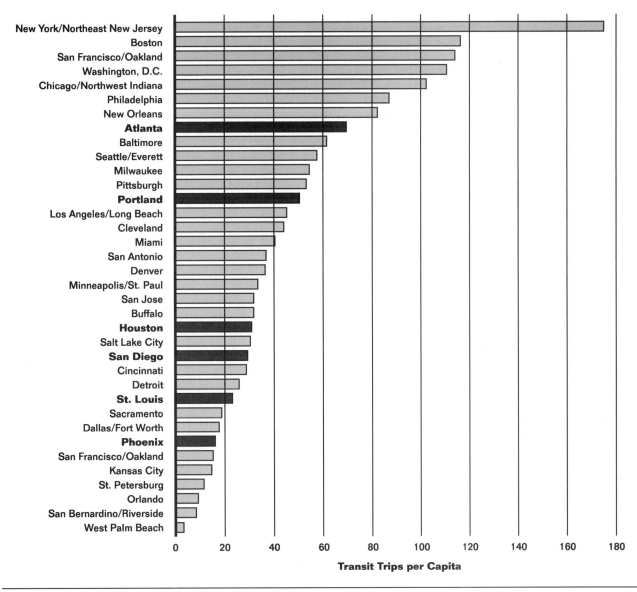

Transit Trips per Capita

Source: 1991 Transit Profiles (Federal Transit Administration).

region barely made the list of the top ten most congested cities.

- Chicago, with its extensive system of bus and rail transit services, ranked near the bottom in the amount of daily driving per capita. However, the freeway system also ranked near the bottom, producing a high overall congestion rating.

Comparing density and travel for different regions shows that while there is some tendency for residents of more dense regions to use transit more and drive less, the statistical case is not strong.[17] Factors other than density apparently have more effect on transit

use and driving. In some regions, especially Atlanta, levels of driving and transit use both are high, while in others, like Phoenix, they both are low. Transit ridership levels in Dallas and Sacramento are at about the same level, but residents of Dallas drive more (the daily per capita miles driven is 23 percent higher). On the other hand, residents of San Francisco and Phoenix have similar driving habits—about 21 miles daily—but San Francisco's transit ridership level is ten times higher than transit use in Phoenix.

Using San Francisco Bay Area data, Greg Harvey and Elizabeth Deakin suggest that VMT per capita declines steeply as density begins to increase, with the marginal rate of decline diminishing above 40 persons per acre. Income appears to account for only a small portion of this effect.[18] A Bay Area study by John Holtzclaw suggests that doubling density from a suburban level to a level equal to that in San Francisco city neighborhoods reduces per capita VMT by 25 to 30 percent.[19] ULI research has estimated that increasing urban densities to a roughly comparable level in other U.S. regions would produce an 18 percent reduction in per capita VMT, but would also increase total VMT by 64 percent.

Current research on density and driving at the national level is limited. Perhaps the most recognized data are from a survey by Jeffrey Kenworthy and Peter Newman covering 32 major cities in the United States and abroad. This study found that density has a significant impact on shifting travel away from cars to transit. It concluded that doubling the density of a city reduces per capita gasoline consumption by 25 to 30 percent.[20]

The wide diversity among regions even for the limited range of transportation and growth factors presented here suggests that each region will need to find its own solutions. However, a community seeking to learn lessons from the experiences of other regions can put comparative evaluations to good use. The Metro planning department in Portland, Oregon, undertook such a study and discovered, contrary to the conventional wisdom, that Los Angeles, with its high densities, low road and freeway mileage, and nearly average travel times, represents a pattern to be emulated rather than a future to be avoided.[21]

Transportation and land use are highly charged, often emotional issues and people's approach to them is usually subjective. In this book, we use analytical methods wherever possible in an effort to bring more objectivity to the issues and make the lessons learned more transferable. However, we recognize that gridlock is not rocket science.

Notes

1. Joel Garreau, *Edge City: Life on the New Frontier* (New York: Doubleday, 1991).

2. Robert Cervero, *Suburban Gridlock* (New Brunswick, New Jersey: Center for Urban Policy Research, 1985), p. 39.

3. Vehicles are driven by people. Measures of capacity are based on average driving performance. They relate how well vehicle operators—commuters, vacationers, truckers, and bus drivers—can pass through an intersection or merge onto a freeway. If, for example, drivers were able to start immediately after a signal turned green and travel though the intersection bumper to bumper, that intersection's capacity would increase. Although not necessarily a safe way to drive, this possibility illustrates the human aspect of capacity measures.

4. Alan E. Pisarski, *New Perspectives in Commuting* (Washington, D.C.: U.S. Department of Transportation, 1992), p. 91.

5. Peter Gordon and Harry Richardson, *Geographic Factors Explaining Work Trip Length Changes*, 1990 Nationwide Personal Transportation Survey: Special Reports on Trip and Vehicle Attributes (Washington, D.C.: U.S. Department of Transportation, 1995), p. 2-14.

6. Gopal Ahluwalia, *Profile of the New Home Buyer* (Washington, D.C.: National Association of Home Builders, 1991), p. 36. While the NAHB database is much smaller than national survey databases, it appears to be reasonable and consistent over time.

7. National Association of Home Builders, *What Today's Home Buyers Want*, executive summary (Washington, D.C., National Association of Home Builders, 1996), p. 3.

8. Patricia Hu and Jennifer Young, *1990 NPTS Databook*, Nationwide Personal Transportation Survey (Oak Ridge, Tennessee: Center for Transportation Analysis, Oak Ridge National Laboratory, for the Office of Highway Information Management, Federal Highway Administration, 1993), p. 4-180.

9. Rolf R. Schmitt, *Transportation Statistics Annual Report: 1994* (Washington, D.C.: U.S. Department of Transportation, 1994), pp. 102, 118.

10. U.S. Department of Transportation, *1995 Status of the Nation's Surface Transportation System: Conditions and Performance*, report to Congress (Washington, D.C.: U.S. Department of Transportation, 1995), pp. 68, 115–116.

11. Ibid., pp. 116, 137, 163.

12. National Council on Public Works Improvement, *Fragile Foundations: A Report on the Nation's Public Works* (Washington, D.C.: U.S. Government Printing Office, 1988), p. 8.

13. Bay Area Economic Forum, *Market-Based Solutions to the Transportation Crisis*, executive summary (San Francisco: Bay Area Economic Forum, 1990), p. 2.

14. Anthony Downs, *New Visions for Metropolitan America* (Washington, D.C.: The Brookings Institution, 1994), pp. 6–15.

15. Mary K. Teets, *1993 Highway Statistics* (Washington, D.C.: U.S. Government Printing Office, 1994), p. V-72.

16. Alan E. Pisarski, *Commuting in America*, 2nd ed. (Washington, D.C.: Eno Transportation Foundation, 1996), p. 63.

17. A statistical regression of density shows that density "explains" 15 percent of the regional variation in daily driving and 26 percent of the variation in per capita transit use. See Robert T. Dunphy and Kimberly Fisher, "Transportation, Congestion, and Density: New Insights," *Transportation Research Record 1552* (Washington, D.C.: Transportation Research Board, 1996), pp. 89–96.

18. Greg Harvey and Elizabeth Deakin, *Toward Improved Transportation Modeling Practice*, rev. draft (Washington, D.C.: National Association of Regional Councils, 1991).

19. John Holtzclaw, "Using Residential Patterns and Transit to Decrease Automobile Dependency and Cost" (paper prepared for California Home Energy Ration Systems by the National Resources Defense Council, San Francisco, 1994). This study resolved the difficulty of estimating the travel distances of residents of different neighborhoods by using odometer readings checked in emissions inspections.

20. Jeffrey Kenworthy and Peter Newman, *Cities and Automobile Dependency* (Brookfield, Vermont: Gower Publishing Company, 1990).

21. Sonny Condor, *Metro Measured,* Metro Occasional Paper Series, no. 3 (Portland, Oregon: Metro Planning Department Data Resource Center, 1994), p. 7.

Chapter 2

How Projects Are Developed

The basic controversy is about power and money, specifically whether technical or political criteria shall determine the use of federal highway funds.

— Jonathan L. Gifford, George Mason University

Anyone who hopes to improve the relationship between development and transportation facilities needs to understand why and how developers undertake real estate projects and why and how transportation advocates undertake the development or expansion of transportation projects. In both cases projects are being developed. To call the people who put together real estate projects and the people who put together transportation projects by the same name—developers—may seem outlandish to those who think of the former as rich, high-flying entrepreneurs doing deals and the latter as nerdy engineers pushing projects through reluctant bureaucracies. Neither of these stereotypes is valid and the two development processes have much in common. This chapter presents an overview of each and draws parallels wherever possible.

The development process is probably the most glamorous and controversial aspect of both real estate and transportation, but it is only one part. Consider the economy. The contribution of real estate capital to the nation's gross domestic product (GDP) in 1992 was estimated at $1,223 trillion, 20 percent of GDP and double other capital contributions.[1] Consider residential real estate. The 21 million housing units that were built between 1980 and 1994 represented only 24 percent of total housing units existing in 1980, and some of the new construction was to replace existing units.[2] Consider transportation improvements. The annual average of 15,000 miles of highways under construction between 1990 and 1994 represented less than 0.5 percent of the miles of public roads in the United States, and only 530 miles of new routes were included in this highway construction.[3]

Real estate development is often performed by individuals and entities that are not, strictly speaking, developers—corporations, institutions, governments, landowners, and so forth. Many types of entities are similarly involved with the management of real estate, especially commercial real estate. Real estate commonly accounts for 20 to 25 percent of the assets of corporations not involved in the finance, insurance, or real estate sectors, and it can account for as much as 30 to 50 percent of the assets of some corporations. Many individuals own property for investment (income and appreciation) purposes. Managing commercial real estate is a service business that involves operating, maintaining, and improving property. During the last real estate recession, many developers could not continue developing and turned to real estate management as a way to keep their doors open, while full-service real estate companies with their own management departments put more emphasis on the management side of the business.

Companies that include both a development group and a management group generally operate them with separate teams, and often the two groups have different capitalization structures and financial participation schedules in which the development principals are required to have significant ownership stakes.

For major transportation projects, the structure of the transportation agency is also important. While the project developers cannot be given an ownership stake, agencies tend to put their most able communicators on the front line for controversial projects. The compensation earned by persons in charge of the development team for transportation projects and that earned by the traffic planners and engineers who are in charge of operating and maintaining transportation projects are probably about the same. In real estate firms, in sharp contrast, compensation for development managers is normally much higher than compensation for property managers.

Developing Real Estate

Market fundamentals are the key drivers of development. Developers seek out opportunities to participate in markets with declining vacancy rates and increasing rents. A project begins with an idea. The concept stage is often the most difficult stage in real estate development. It can occupy 20 to 30 percent of the total time spent on a project. Ideas are generated in many different ways.

A site may be searching for a use. Developers often come upon such sites. They may be contacted by the owners of a parcel, who, for one reason or another, want it to be developed. An axiom of real estate is that the worst reason for developing a property is because you own it.[4] It is important to find a use that can be economically developed and will be supported by the market. Figure 2-1 shows an analysis for a site searching for a use.

Or developers might find a use looking for a site. Frequently, a company or institution that wishes to expand contacts a developer or real estate broker. A user-driven development process is most successful when the developer puts together a development program based on the needs of the likely users and suitable investment criteria and then acquires land most likely to advance that development program.[5]

A third way in which real estate development is initiated is capital in search of investment. Institu-

Figure 2-1
Analysis Process: In Search of a Use(s) for a Site

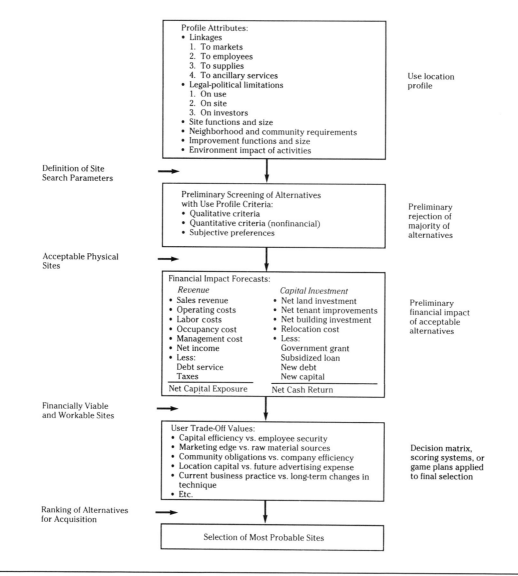

Source: Graaskamp on Real Estate (ULI, 1991).

tional investors such as pension funds and REITs seeking development opportunities not only offer a market for developers, but also an important funding source.

No matter the source of the concept—an available site, a company wanting space, or money seeking a return—the development process is more or less the same, and illustrated in Figure 2-2.

Figure 2-2
The Eight-Stage Model of Real Estate Development

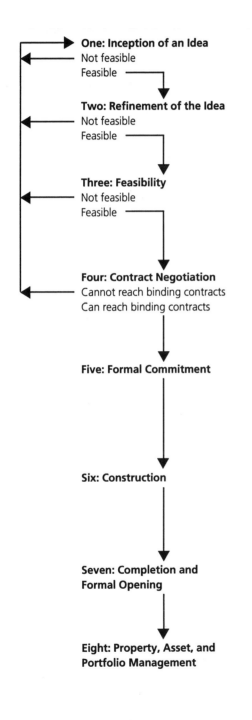

One: Inception of an Idea
Not feasible
Feasible

Developer with background knowledge of the market looks for needs to fill, sees possibilities, has a dozen ideas, does quick feasibility tests in his head (legal, physical, financial).

Two: Refinement of the Idea
Not feasible
Feasible

Developer finds a specific site for the idea; looks for physical feasibility; talks with prospective tenants, owners, lenders, partners, professionals; settles on a tentative design; options the land if the idea looks good.

Three: Feasibility
Not feasible
Feasible

Developer conducts or commissions formal market study to estimate market absorption and capture rates, conducts or commissions feasibility study comparing estimated value of project to cost, processes plans through government agencies.

Four: Contract Negotiation
Cannot reach binding contracts
Can reach binding contracts

Developer decides on final design based on what market study says users want and will pay for. Contracts are negotiated. Developer gets loan commitment in writing, decides on general contractor, determines general rent or sales requirements, obtains permits from local government.

Five: Formal Commitment

Contracts, often contingent on each other, are signed. Developer may have all signed at once: joint venture agreement, construction loan agreement and permanent loan commitment, construction contract, exercise of land purchase option, purchase of insurance, and prelease agreements.

Six: Construction

Developer switches to formal accounting system, seeking to keep all costs within budget. Developer approves changes suggested by marketing professionals and development team, resolves construction disputes, signs checks, keeps work on schedule, brings in operating staff as needed.

Seven: Completion and Formal Opening

Developer brings in full-time operating staff, increases advertising. City approves occupancy, utilities are connected, tenants move in. Construction loan is taken out, and permanent loan is closed.

Eight: Property, Asset, and Portfolio Management

Owners oversee property management, including re-leasing; longer-term owners oversee reconfiguring, remodeling, remarketing space as necessary to extend economic life and enhance performance of asset; corporate management of fixed assets and considerations regarding investors' portfolios come into play.

Source: Real Estate Development: Principles and Process (ULI, 1996).

17

The Preliminary Pro Forma

The developer's first screening of a concept looks at the financial return via an informal, back-of-the-envelope pro forma. For an office or retail project, the developer estimates market rents and roughly calculates what it would cost to build and operate the project. The difference between the projected income stream and operating costs represents the net operating income. Using a standard ratio for the type of project under consideration, a ratio that reflects the market demand for that type of product as well as prevailing interest rates, the developer capitalizes the income stream and compares the resulting net value to the land and development cost estimates. If value exceeds cost by an appropriate rate of return, the idea lives to the next stage. If not, it goes back to the drawing board.

Evaluating Feasibility

The next stage is the refinement of the idea. Ideas that look promising are subjected to successively more detailed evaluation. Costs begin to escalate as design studies, market research, and engineering studies are initiated, so the developer needs more assurance that the project will be successful. At this stage, the developer will be talking with prospective tenants, lenders, and partners. One means of minimizing risk is to settle on a tentative design and take an option on a site.

The design of transportation access is an important consideration in this early phase of development, and it involves many factors such as zoning, soil conditions, utilities planning, and even local attitudes toward development. Transportation improvements and impact fees for transportation must be incorporated in the project budget.

As the idea is continually refined, the developer can still revise the project based on market considerations, changing the product, for example, from industrial to residential. The management structure may also change, with partners being dropped or added. A public partner, such as a city or transportation agency, may be taken on to share development costs and risks as long as it does not impose unacceptable conditions or cause delays.

As the feasibility stage progresses, the developer's involvement in the deal becomes deeper. While backing out is still possible, the further along the project is the higher the sunken costs—financial and emotional—that will have to be written off. If the developer's formal market study shows adequate absorption trends and capture rates, and if the project is still deemed feasible, negotiations get serious and design changes are made to reflect user needs.

Implementation

Once final contracts are signed, construction and lease-up or sales can begin. The various commitments that must be obtained are often contingent on each other. Financing commitments are often contingent on the developer's ability to lock up lease agreements for a certain amount of the space. Funds for building a house may not be released until a buyer signs a contract.

Navigating the initial approval process and, for commercial projects, meeting preleasing commitments pose the highest risks for the developer in the implementation phase. But there are other considerable risks in implementation as well. Unanticipated problems in construction may arise. The expected users may not materialize. Competitive projects may be developed nearby. It is not unusual for the developer to lose the project in the implementation stage. A developer's tradition holds that in large-scale community development projects only the third developer makes money (after the first two have written down many of the early costs).

Management

Developers, if they own the projects they develop, either sell them or retain them in their portfolios. Buyers may be users or entities that wish to acquire investment properties after the high-risk development phase is complete. Whether these projects are single-family houses, build-to-suit industrial or office buildings, or developments such as shopping centers, apartment buildings, hotels, or multitenant office buildings from which operational returns are expected—all are presumably long-lived and require long-term management.

The management of investment property involves day-to-day maintenance and repairs as well as releasing as users leave. Over time, it may involve expanding, remodeling, or reconfiguring the asset, turning a community shopping center, for example, into a high-end fashion center.

Also over time, transportation changes can increase or reduce the value of a property. Highway or transit improvements that significantly improve access can make a property more valuable. Changes that make the property more difficult to reach—such as the closing of a highway ramp or the elimination of a median crossing—can depress its value.

Developing Major Transportation Projects

Developers and others involved in the development process often are put off by the terminology and procedures of transportation planning and project development. They tend to view transportation planners as engineers-with-an-attitude in search of some transportation holy grail. They find it difficult to penetrate the arcane, quantified language of travel forecasting and transportation systems planning. But there is a great interest among federal officials in making transportation more inclusive, and some states share this attitude as well. The process of developing major transportation projects has changed a great deal in the last several years.

The ISTEA Environment

Federal funding can pay up to 80 percent of the costs of major highway and transit projects, so their development is heavily influenced by federal rules and guidelines. Chief among current guidelines is ISTEA —the Intermodal Surface Transportation Efficiency Act of 1991. ISTEA reshaped funding, intergovernmental relationships, and the highway and transit relationship, according to the promise by its chief sponsor, Senator Daniel Patrick Moynihan of New York, of "a new ball game in transportation policy."

More revolutionary still was the linking of ISTEA to two environmental laws—the 1990 Clean Air Act Amendments and the Comprehensive National Energy Policy Act of 1992—to produce a remarkable triad of federal legislation that deals with the complex interaction between transportation, air quality, and energy, and makes very subtle forays into the touchy subject of land use.

ISTEA was great news to environmental and planning groups, local officials, and transit supporters who did not like the way the transportation funding game had been played. But federal and state transportation officials and highway advocates who were no longer as well received in the halls of Congress considered it a major disappointment.[6] The two sides engaged in battles over the implementing regulations and the battles will no doubt intensify during the debate over the act's reauthorization in 1997.

Developers tended to think that ISTEA was a step in the right direction. Transportation consultant Daniel Brame of Kimley Horne's Orlando office, who recently moved to Portland, says that at first developers in Florida were optimistic, thinking "that ISTEA would provide more money and better and more reliable schedules when it comes to implementing roadway projects." But, he continues, they found that it did not make life easier: "The new process is more complex and involves more players, more political agendas. A developer that needs transportation capacity at a certain time must continually monitor and stay involved in the process."[7]

ISTEA clearly signals a broader view of how transportation projects fit into the urban environment, to replace a sometimes narrow view of the transportation problem and project fixes. Before ISTEA, says Sheldon Edner of the Federal Highway Administration (FHWA), "project development was seen as a production problem, like manufacturing widgets. Now that planning counts more, product development should reflect these changes. In the past, how to fix a bridge was the focus and we studied a few limited options. Under the new guidelines, the issue is bigger—not how to fix the bridge, but what is the transportation problem we are trying to solve."[8] Unfortunately, while the broader view should lead to better projects, it will also complicate and possibly draw out the project development process.

Federal guidelines prescribe some specific planning procedures. Bureaucratic technocrats have been criticized sometimes for developing a project selection process that limits political influence. One researcher, reviewing the debate that led to the adoption by Congress in 1995 of the national highway system, observes:

> Not surprisingly, the basic controversy is about power and money, specifically whether technical or political criteria shall determine the use of federal highway funds. The highway bureaucracy is bred with a very strong streak of progressivism, which manifests itself in the desire to allocate funds on the basis of objective, technical criteria, rather than on the basis of politics. Congress and other political bodies are somewhat differently disposed.
>
> Systems of highways, designated on the basis of technical criteria, have been an effective first line of defense from political tampering for 75 years. Recently, however, as the construction of the Interstate system has wound down, political forces have exercised increasing influence over the disposition of highway funds, largely through the use of congressionally earmarked demonstration projects, which the transportation community views very unfavorably.[9]

While the planning requirements for highway projects and transit projects are similar, the federal government has exercised much more control over the development of transit projects. Harold Peaks, team leader for community impacts in FHWA, points out that the federal governments does not initiate highway projects. That role typically belongs to state transportation agencies, which have been traditionally the key players in the development of regional highway networks. These regional networks are based on a strong core of state highways. Yearly funding allocation is based on a national formula, and states are given great freedom to select individual projects. At one time, highway projects needed to satisfy cost-benefit standards, but that requirement has been dropped. While FHWA does not tell a state which projects to undertake or not undertake, it does, says Peaks, "stand accountable for the public dollar." The normally hands-off FHWA might try to influence certain highway decisions.[10]

Federal transit investments have been subjected to much greater scrutiny and control, reflecting a view that highways play an important role in the national economy (the interstate commerce function) while transit is more of a local concern. The different treatment of the two types of projects also reflects funding differences, with federal highway funds coming exclusively from user taxes and federal transit funds coming from a mix of highway-user taxes and general revenues.

Major transit proposals were required to comply with a rigorous set of criteria. Federal transit funds were limited, and the intent was to allocate them cost-effectively. During the 1980s, federal transit funding was a political football. Believing that transit is not a federal responsibility, the Reagan administration eliminated all funding for new rail projects. Not sharing that view, Congress restored the funding and earmarked it for individual projects in order to preclude the head of the U.S. Urban Mass Transit Administration from having a say. One senior official who at the time was responsible for reviewing the cost-effectiveness of transit projects quipped: "If they cannot get their money through the FTA [Federal Transit Administration] window, they go to the Mark Hatfield [Senator from Oregon] window." Since ISTEA, there has been an effort to provide more even treatment of highway and transit projects, as described below under "The Major Investment Study."

Local and State Roles

The participation of metropolitan planning organizations (MPOs) has been a requirement of federal transportation funding since the mid-1960s, and one that many states chafe under. Designated by the federal government to ensure that the projects it funds address a range of regional issues, MPOs can be established regional planning organizations, councils of governments, or, in some smaller regions, units of a state department of transportation (DOT). Generally, they are made up of local governments and state DOTs participate in their work. MPOs are as close as the federal government has been able to come to the establishment of a metropolitan land use planning function. They have been likened to "a party where everyone hates the host, and wishes they weren't there, but aren't allowed to leave."[11]

State transportation agencies are major road builders. They are responsible for 20 percent of total public street and road mileage, and this one-fifth of all roads carries 70 percent of all highway travel. State funding for highways has represented about half of all government highway spending since the 1960s. With the exception of the 1976–1986 period, state and local governments have consistently supplied more than 50 percent of highway investments. In 1993, government highway spending was shared as follows: state governments, 53 percent; local governments, 26 percent; and the federal government, 21 percent. Before ISTEA, federal spending was restricted to capital investment, and states were responsible for road maintenance.[12] States are also making an increasing contribution to transit, and they currently outspend the federal government in the area of transit assistance. Of the $88 billion in public sector spending for highways in 1993, 61 percent came from user charges, mostly motor fuel taxes, with the remainder coming from other sources such as property taxes, general funds, bonds, exactions, development fees, and special district assessments. (Note that this represents only public funds actually spent on highways. States collected $10 billion more in user revenues than they spent on highways.) User fees account for a much larger share of federal and state spending on highways, 87 percent and 72 percent, respectively. Of $23 billion spent on transit in 1993, 31 percent was financed from users, with the remainder coming from dedicated taxes, general funds, and motor fuel taxes.

Traditionally practitioners of the golden rule (them with the gold rules) in project selection, states have adopted a more collaborative approach, whether they like it or not, under ISTEA. Florida has 25 urbanized areas and is highly supportive of MPOs. According to Doug McLeod of the Florida Department of Transportation, the state's MPOs are among the most influential in the United States; the DOT sets up a state work program and participates in planning, but most major priorities are set at the metropolitan level. The state's flexible approach carries over to nonfederal transportation projects, which can help expedite the development of transportation facilities that are needed to support (wanted) real estate projects.[13]

Recognizing that there is not sufficient funding for all valuable projects, the state of Washington has adopted a strategic planning method based on five goals for the state's transportation program: preserve and maintain the existing system; improve safety; encourage mobility choices; support economic growth; and foster environmental responsibility. A scoring system makes it possible to rank proposed projects by their contribution to the state's goals. Geographic equity is obtained by selecting the highest ranked projects within each of the state DOT's six regions, making it possible, for example, to fund a project in Spokane that might not make the list in Seattle.

Figure 2-3

The Elements of the Regional Transportation Planning Process

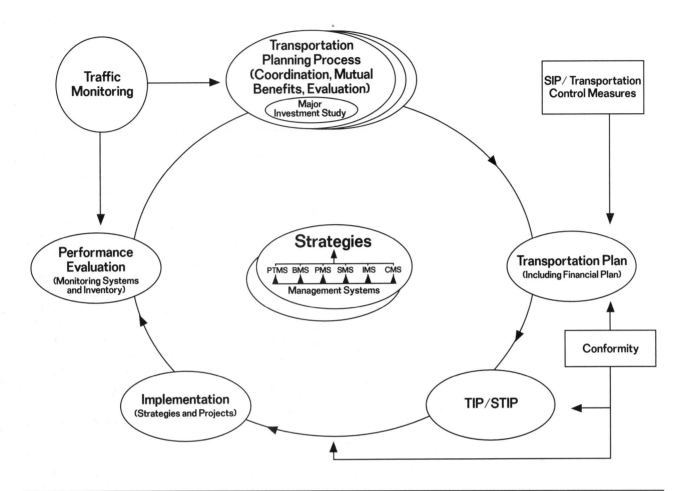

Source: *A Guide to Metropolitan Transportation Planning under ISTEA* (U.S. Department of Transportation, no. 95–967).

Developing transportation projects in Washington is a complex process, says Charles Howard of the state DOT. The state's growth management legislation requires that transportation investments be consistent with local plans. "We all point at each other and say: 'You have to be consistent.'" Growth management policy is another reason for the state DOT to work within the MPO process.

All parties are trying to make the process work and a good example of how well it has worked to meet business growth needs is an interchange that was added to support a new Intel plant in Dupont, a town north of Seattle. The process worked because Intel's expansion had been anticipated, and the project had been in the state transportation plan for six years.[14]

Wisconsin has also stepped back from old ways of transportation planning. In response to a number of emerging issues, but most of all ISTEA, the state

developed a comprehensive 25-year vision and comprehensively multimodal plan for transportation in the 21st century. Among the key features of the plan are a commitment to assist MPOs in urban planning, the setting of goals for intercity freight and passenger transport, an emphasis on state financial support of transportation projects, and perhaps trendsetting provisions for conducting environmental evaluations at the system level.

Long-Range Plans

Federal rules require MPOs to develop 20-year transportation plans that identify the facilities—including major roadways, railroad stations, port terminals, airports, and others—that will be needed in an integrated, intermodal regional transportation system. These plans should include an assessment of the short- and long-term actions that will be necessary for developing and maintaining these facilities. The regulations list 16 specific planning factors, grouped into three areas:

Mobility and Access for People and Goods
- Inclusion of all projects, even those developed without federal funds;
- International border crossings and access to critical areas;
- Road connectivity of metropolitan area to broader market;
- Enhancement of efficient freight movement; and
- Transit services and use.

System Performance and Preservation
- Congestion relief and prevention;
- Improved management of existing facilities;
- Identifying needs through the implementation of management systems;
- Preservation of rights-of-way; and
- The use of life-cycle costs in designing bridges, tunnels, and pavement.

Environmental Quality and Quality of Life
- Social, economic, energy, and environmental impacts of transportation;
- Consistency of planning with energy conservation;
- Relationships between transportation and land use planning;
- Spending on transportation enhancements;
- Improvements in transit system security; and
- Tourism and recreation.

In reaction to concerns that transportation planning has focused too much on investment in new projects and not enough on ways of making existing assets more productive, the new federal regulations place heavy emphasis on improved management. Transportation agencies must put systems in place for managing pavement, bridge, and public transportation assets. They must prepare reports on the performance of entire transportation systems, including safety, congestion management, and intermodal connections.

Regional transportation planning also must abide by some other key federal mandates. It must be in conformance with regional air quality plans. It must involve the public. Major projects require major investment studies (see "The Major Investment Study" section in this chapter). And, transportation planning must incorporate a financial plan that is reasonably attainable. The requirement for a realistic financial plan addresses one of the most frequent criticisms of metropolitan transportation planning, namely, that it can be unrealistic. Transportation planners have been known to rely on an easy way to deal with tough projects: put them in the plan, but never build them.

The regional transportation planning process incorporating the interaction all of these elements is shown in Figure 2-3. Federal regulations require that MPOs issue a periodic action plan, which is called the Transportation Improvement Program.

Transportation Improvement Program

No federal project can be implemented unless it is included in the Transportation Improvement Program (TIP). See Figure 2-4. The TIP has at least a three-year horizon, and it must be updated every two years. The projects it includes must be consistent with the long-range regional transportation plan. The TIP is a strategic management tool for accomplishing the objectives of the plan and a link between the planning agency and the operating agencies that actually implement the plan.

As noted, conformity with air quality regulations is required for long-range transportation plans. New or amended TIPs must conform with the approved state implementation plan (SIP) for air quality. (Conformity should be straightforward at this stage.) And they must give priority to the SIP's transportation control measures (TCMs) for any areas not meeting air quality standards. The TIP must be approved

Figure 2-4

From Transportation Plan to Project Development

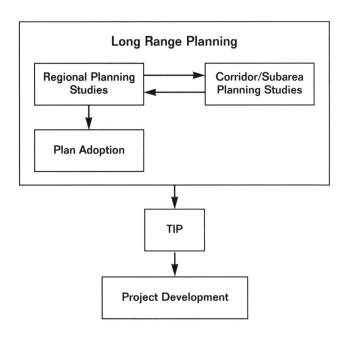

Source: A Guide to Metropolitan Transportation Planning under ISTEA (U.S. Department of Transportation, no. 95–967).

by the MPO and the state governor, and a conformity determination must be made by the MPO and approved by FHWA and FTA.

The Major Investment Study

Regional planning can be a fairly unexciting process until a major project is proposed and opponents come out of the woodwork, which they usually do almost regardless of the project. Before ISTEA, reviews of costly, controversial projects usually took place outside the planning process after key policy decisions had been made. Studies tended to focus on a single mode of transportation. Major transit proposals were put through special corridor studies called alternatives analysis. Highway proposals underwent environmental studies at the project level. These studies often considered alternatives at a high level of detail, adding to the cost and time of project development.

In 1993, new regulations issued by FHWA and FTA to carry out the planning provisions of ISTEA created the major investment study (MIS) process. In the words of Donald Emerson of the FTA Office of Planning, this involves state and local planners working together "to define transportation problems and to evaluate alternative solutions during the planning process, before plans are adopted." The result is a broader consideration of alternatives, which often include multimodal strategies.[15] At the same time, FHWA and FTA are shifting the federal role in the planning process from prescription and regulations to partnering and technical assistance.

Embedding major investment studies early in the planning process, it is hoped, will lead planners to consider more creative and mixed solutions. For example, a new highway could be designed with fewer lanes if its construction were paired with improvements on existing facilities. Or a cooperatively planned mix of transit and highway improvements could enhance mobility more than a single-mode solution. Already, there has emerged much greater cooperation between highway and transit program senior staff in the federal agencies in working out planning regulations and dealing with MPOs.

While the planning procedures for major investments have become more flexible, there are still significant differences in the approval process for major highway and transit projects. The allocation of federal highway funds to states is by formula, so that once a state has decided to proceed with a project, federal approval is virtually guaranteed. Transit funding is by both formula and discretionary programs, and major investments usually involve discretionary funds, which means that the projects must comply with various criteria and be approved by the FTA administrator. Furthermore, Congress's practice of earmarking transit funds removes much of the discretion that can be exercised by FTA and makes congressional appropriations a part of the process that local project sponsors must go through to get discretionary transit funds.

Some key themes of the ISTEA initiative—collaborative decision making, proactive public involvement, early consideration of environmental factors, regulatory streamlining, and consideration of broad multimodal project alternatives—have been included in some early major investment studies, three of which are described next.[16] While these three studies are not necessarily models, the federal agencies view them as "poster children" for this new process.

U.S. 78 Corridor Study: Atlanta. U.S. 78 is a major east/west route serving the Atlanta region's fast-growing northeast suburbs—DeKalb and Gwinnett counties—that has severe congestion and safety problems. The route is part freeway, part six-lane surface arterial street with reversible lanes through a busy commercial area, and part rural road with two lanes in each direction and a center turn lane. Various proposals for localized improvements never gained the support of the local community or transportation agencies. The Atlanta Regional Council began an MIS in 1994, and this evaluation of issues and solutions shifted the focus of thinking from localized improvements to broader mobility issues.

An extensive public involvement program that reached out to government agencies and citizens of the affected communities included an impressive mailing list, frequent ads in local newspapers, a hot line, a video, and an interactive CD-ROM presentation. A conscious effort to "get out of the box" or take a fresh look at the situation led to a much wider range of concepts being considered than theretofore and, according to the participants, enabled the council to recommend a strategy that departs from previous plans. The MIS calls for a combination of highway improvements: barrier-separated express lanes and the upgrading of U.S. 78 to an expressway. Transit extensions will be considered, facilities for pedestrians and cyclists will be emphasized, and landscaping projects will improve the look of the corridor.

Pocatello/Chubbuck I-15/I-86 Corridor Study: Idaho. The Pocatello/Chubbuck metropolitan area is a small (55,000 population) but growing region in which the two major interstates have developed significant congestion problems. The Idaho Department of Transportation had received $10 million in 1991 to be used for relieving congestion in the I-15/I-86 corridor, but it had made no significant progress by 1994, when the Bannock Planning Organization requested that the state agency use the MIS process to clearly define a solution and generate public support.

A decision was made to engage in proactive and early outreach, which would not only comply with the spirit of the process but also provide needed support during the environmental approval process. The planners used a newspaper survey and listening posts—information booths located in public places and staffed by planners who were able to answer questions—to refine the range of study options. This public involvement was instrumental in the selection of a project, which was to improve two interchanges and widen I-15. Because of the collaborative planning process and public involvement, the project's environmental review requirements can most likely be addressed with an environmental assessment rather than a full environmental impact statement, for a potential savings of six months from the two years normally needed to prepare an EIS.

State Highway 152 Corridor Study: Oklahoma City. SH-152 is a two-lane road southwest of Oklahoma City that provides access to adjacent residential, commercial, and industrial areas, including the FAA Training Center and the Will Rogers World Airport. The need for safety and capacity improvements has been acknowledged for years, and, in fact, an environmental assessment was approved over a decade ago for widening the road and realigning part of it. Thinking that the long-term development of the corridor might be enhanced by connecting SH-152 to Airport Road, a divided, controlled-access facility, the city of Oklahoma City assumed responsibility as lead agency for an MIS soon after FHWA and FTA created the MIS process. This revisit of the corridor plans prompted a broader consideration of the corridor's role in serving local travel, freight movement, and access to the airport and FAA center. The selected investment strategy is expected to do a better job serving commuters as well as airport passengers and freight.

The MIS process has changed the nature of planning. Whether it will result in better decisions is still an open question. One disadvantage of an open, flexible, participatory planning process is that it can take longer and keep alive options that do not make good technical sense. Of course, the best solution is of little value if it does not have public support. As Sheldon Edner points out: "Investing in planning and project development is cheap, especially if it solves problems in advance through early anticipation of undesirable impacts and the identification of the most effective solution."[17]

Exerting Influence

Knowing how to influence the planning process for major transportation projects can make a big difference. The new federal interest in making transportation planning more inclusive has opened up new avenues for outsider influence. Now it is easier for developers, planners, and others involved in growth and development issues to have an impact. Challenging the data, bringing money to the table, and getting involved are three good ways of making a impact

Transportation planning is a data-hungry process. Forecasts of travel demand rely on forecasts of growth and land use, and land use must be quantified to levels that can often infuriate planners. Forecasts of traffic that will be generated by different land uses, such as shopping centers, often are dismissed as technical exercises needed to feed the computer, but they should be scrutinized closely by people with a stake in how development affects traffic. Some developers have become unwitting participants in the data game as they comply with requirements to estimate the traffic impacts of not only their projects but of others expected to come on line in the study area or transportation corridor.

Transportation planners reduce alternative visions of a region's future to a set of numbers for each defined zone of transportation analysis. (These zones are usually smaller than Census tracts.) Critics have pointed out that putting such precise numbers on such fuzzy futures is risky business, but that is the way transportation decisions are made. The easiest way to change plans is to change the inputs, either by proposing plausible alternative futures or by pointing out errors and inconsistencies in the transportation plan's quantification of the future. Gregory Benz of Parsons Brinckerhoff's Baltimore office has directed complex transportation studies for developers and public agencies. He suggests proactive efforts to change the forecasts used by transportation planners, saying "by the time the consultants are brought in, land use is a given, and we are in a reactive mode."[18]

Money is a second avenue of influence. Despite the image of transportation planning as an unbiased, scientific search for optimal solutions, the financial bottom line is more important than ever, especially with the new federal requirement that plans be financially constrained. James Bednar, who previously directed project development for FHWA and now works for the consulting firm of Parsons Brinckerhoff in Columbus, Ohio, reports that money talks, and many senior federal and state officials and other transportation consultants agree with this assessment. Private sector or local government funding contributions to a state project can influence when it will be undertaken, often advancing it ahead of other state-funded projects.

The third avenue is to get involved. Individuals are welcome in the open transportation planning process, and they can easily gain access to information through meetings and newsletters. Some outsiders have discovered this. Bicycle advocacy groups, for example, have become actively involved in MPO activities and in many cases they have successfully advanced their agenda for bicycle facilities.

Notes

1. ULI–the Urban Land Institute, *America's Real Estate* (Washington, D.C.: ULI–the Urban Land Institute, forthcoming in 1997).

2. U.S. Bureau of the Census, *Statistical Abstract of the United States: 1995* (Washington, D.C.: U.S. Government Printing Office, 1995), pp. 728, 733.

3. U.S. Department of Transportation, *Highway Statistics: 1995* and *1990* (Washington, D.C.: U.S. Department of Transportation, 1995 and 1990), p. IV-38 (1995) and p. 124 (1990).

4. Richard B. Peiser and Dean Schwanke, *Professional Real Estate Development* (Chicago: Dearborn Financial Publishing, Inc. and ULI–the Urban Land Institute, 1992).

5. James D. Graaskamp, *Graaskamp on Real Estate*, ed. Stephen P. Jarchow (Washington, D.C.: ULI–the Urban Land Institute, 1991), p. 244.

6. Robert T. Dunphy, "Federal Transportation Policy and Land Use: A New Ball Game?" in *ULI on the Future: Urban Growth* (Washington, D.C.: ULI–the Urban Land Institute, 1994), p. 26.

7. Daniel Brame, interview by author, September 4, 1996.

8. Sheldon Edner, interview by author, March 17, 1996.

9. Jonathan L. Gifford, "Historical Antecedents and Development Impacts of Highways of National Significance: The Conflict between Technical and Political Criteria" (paper presented at the annual meeting of the Transportation Research Board, Washington, D.C., January 13, 1991).

10. Harold Peaks, interview by author, June 9, 1996.

11. Robert C. Brannan, Ethan Seltzer, and Michael A. Wert, "Coordinating Portland's Urban Growth Plan and the Western Bypass Study," in *ITE 1991 Compendium of Technical Papers* (Washington, D.C.: Institute of Traffic Engineers, 1991), p. 69.

12. U.S. Department of Transportation, *1995 Status of the Nation's Surface Transportation System: Conditions and Performance,* report to Congress (Washington, D.C.: U.S. Department of Transportation, 1995), pp. 79–81.

13. Doug McLeod, interview by author, June 4, 1996.

14. Charles Howard, interview by author, June 3, 1996.

15. Donald Emerson, interview by author, June 14, 1996.

16. U.S. Department of Transportation, *U.S. 78 Corridor Study* (MIS Case Studies, February 1996); *Pocatello/Chubbuck Corridor Transportation Study* (MIS Case Studies, February 1996); and *Oklahoma State Highway 152 Corridor Study* (MIS Case Studies, February 1996).

17. Edner (see note 8).

18. Gregory Benz, interview by author, May 28, 1996.

Chapter 3

Demographics, Changing Preferences, and Travel

The notion that there is an American "love affair" with the automobile is missing the point. [Automobiles] save time, and it is time pressures, particularly on women, that increase personal vehicle use. . . . Decisions regarding household location and mode (of travel) to work are not made frivolously.

— Alan Pisarski, *Commuting in America*

Among the many complex factors influencing travel behavior are the demographic characteristics of the U.S. population: its size, distribution, and composition. Predicting future travel demand requires an understanding of how the number, location, and nature of drivers and passengers will be changing as we go into the 21st century. Population growth, regional shifts, and continued suburbanization are part of the equation. But travel needs also vary by age, sex, income, and household type. Dramatic changes in the workplace will also affect future commuting patterns. How all of these factors come together will determine the need for new road and transit infrastructure for the foreseeable future.

Some transportation policy analysts have hypothesized that in personal travel the United States will soon reach a saturation point. They suggest that time pressures from work and family responsibilities, combined with past increases in the number of licensed drivers and vehicle ownership, will mean few opportunities for further trip growth. In contrast, the results of the 1990 Nationwide Personal Transportation Survey (NPTS) suggest that most Americans are taking more trips and for longer distances, and that this pattern is unlikely to change.[1] (Figure 3-1 shows some key indicators of total travel activity in 1983 and 1990.)

This chapter looks at how the American people and their jobs will change between now and 2010. Its focus is national in scope, but important regional and local differences will be highlighted and examples presented. Demographic trends portend the continued growth of trips made and greater use of personal vehicles. However, travel growth rates will be slowing down, allowing land use planners and traffic engineers to catch up with travel infrastructure needs.

Population Growth

Relative to their impact on future travel patterns and traffic congestion, national population trends do not all point in the same direction. Some trends suggest continued growth in trips, miles traveled, and the use of private automobiles. Others point to a slowdown in travel activity, especially when compared with the high rates of growth seen in the 1970s and 1980s.

Growth in the U.S. population is slowing, but millions of added travelers will still need to be accommodated. Even if travel habits had not changed in the last 30 years, the sheer growth in the number of Americans would have resulted in more vehicle trips and traffic congestion. During the 1960s and 1970s, the United States added roughly 24 million people per decade. The rate of expansion began to slow during the 1980s, as shown in Figure 3-2, dropping below 1 percent per year. Although the first half of the 1990s saw somewhat of a rebound, aggregate growth is expected to stay below 1 percent per year through 2010. By then, the U.S. population is projected to reach nearly 298 million.

Although the rate of growth is leveling off, millions of residents are being added each year. Between 1990 and 1995, the U.S. population rose by nearly 14 million persons. The U.S. Census Bureau projects a net gain of just under 34.9 million people between 1995 and 2010.[2]

Virtually all Americans old enough to drive are already licensed. According to the 1990 NPTS, the number of licensed drivers in the United States increased nearly 28 percent between 1977 and 1990, adding 35 million new drivers. Women accounted for 60 percent of the new drivers, in part because they entered the labor force in larger numbers than

Figure 3-1

Key Travel Indicators for U.S. Urban Areas, 1983 and 1990[1]

	1983	1990	Percent Change
Person Trips (Millions)	139,670	151,070	8.2%
Person-Miles (Millions)	848,827	1,004,367	18.3
Household Vehicle Trips (Millions)	78,834	96,106	21.9
Household Vehicle-Miles (Millions)	515,705	690,847	34.0
Daily Trips per Person, Weekdays	2.75	3.07	11.6
Daily Vehicle Trips per Person, Weekdays	1.62	2.56	58.0
Daily Vehicle Trips per Household, Weekdays	4.21	4.39	4.3
Average Trip Length, Private Vehicles (Miles)	6.98	7.74	10.9
Average Commute Trip Length, Private Vehicles (Miles)	8.00	9.83	22.9

[1]Data are for urban area households and for trips of 75 miles or less.
Source: NPTS Urban Travel Patterns (Federal Highway Administration, June 1994).

ever before.[3] In 1990, 92 percent of age-eligible males and 85 percent of females held licenses.[4]

The rate of growth in licensed drivers began to slow after 1983 for two reasons. The baby boomers were already licensed; and after the baby boom, fewer people each year reached driving age. Moreover, women under age 50 were already obtaining driver's licenses at rates nearly as high as men. (See Figure 3-3.) There is little room for growth in the percentage of Americans who drive. However, the slowdown in the number of additional licensed drivers each year will not, in and of itself, halt the rise in the number of trips or miles traveled.

Figure 3-2

U.S. Population, 1960–2010[1]

	Population[2]	Average Annual Percent Change
1960	179,979,000	1.67%
1970	203,810,000	1.24
1980	227,225,000	1.09
1985	237,924,000	0.92
1990	249,391,000	0.94
1995	262,820,000	1.05
2000	274,634,000	0.88
2005	285,981,000	0.81
2010	297,716,000	0.80

[1]Data from 1995 on are estimates or projections.
[2]Resident population as of July 1.
Source: Population Projections of the United States by Age, Sex, Race, and Hispanic Origin: 1995 to 2050 (U.S. Bureau of the Census, 1996).

In the United States there already is more than one car, van, or light truck for every person age 16 and older. The Federal Highway Administration (FHWA) reported over 146.3 million registered automobiles (including taxis) in the United States in 1993, and another 39.8 million light trucks not used for farm work.[5] That same year, about 173 million Americans held driver's licenses. Among two-adult households, 76 percent had two vehicles in 1990, up from 65 percent in 1983. These facts suggest that vehicle ownership, like licenses, could be reaching the saturation point. What has not peaked yet is the opportunity to drive these vehicles more frequently and for longer distances.

Aging Americans

The addition of nearly 35 million travelers over the next 15 years implies further worsening of local gridlock and the need for significant expansion of the transportation network. But there are good reasons to believe that the travel demands associated with future population growth will be somewhat less than in the past. The most important factor is the aging of the population, as highlighted in Figure 3-4.

Looking ahead to 2010: the number of young adults will return to its 1980 level, from a recent decline. In 1980, young people age 18 to 24 were 13.3 percent of the population and numbered over 30 million. The 1995 estimate shows only 24.9 million persons in this age group, which now constitutes only 9.5 percent of the total population. The 1990 NPTS demonstrated that young men make more trips in an average day than do their older counterparts, although they are somewhat more likely to travel by modes

28

Figure 3-3

Licensing Rates by Gender and Age, 1990

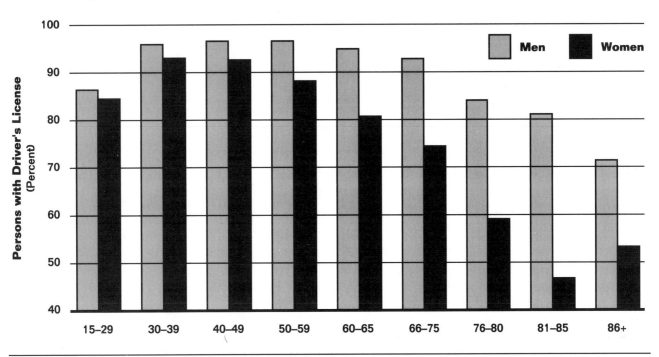

Source: 1990 NPTS: Demographic Special Reports (Federal Highway Administration, 1995).

other than private automobile and, therefore, to drive slightly fewer total miles per day. (See Figure 3-5.) The number of miles driven by teenage boys increased 75 percent between 1969 and 1990, according to the NPTS; for teenage girls, it more than doubled. Increases in teenage driving will be smaller as the population ages over the next decade, and as the proportion of the total population first reaching driving age

declines, although the number of new drivers will continue to rise.

Baby boomers will make up a smaller share of the population by 2010. The boomers (people born between 1946 and 1963) are already in their peak driving years. The sheer size of the baby boom and its affinity for driving were responsible for much of the increase in traffic in the 1970s and 1980s. But the

Figure 3-4

U.S. Population by Age, 1970–2010[1]

	Age Category					
	18–24	**25–34**	**35–54**	**55–64**	**65–74**	**75+**
1970	23,714,000	24,943,000	46,336,000	18,602,000	12,443,000	7,530,000
1980	30,022,000	37,082,000	48,434,000	21,703,000	15,581,000	9,969,000
1990	26,942,000	43,211,000	62,492,000	21,112,000	18,045,000	13,033,000
1995	24,926,000	40,863,000	73,607,000	21,139,000	18,758,000	14,785,000
2010	30,138,000	38,292,000	82,085,000	35,283,000	21,057,000	18,352,000
Percent Change						
1970–1990	14%	73%	35%	13%	45%	73%
1995–2010	21%	–6%	12%	67%	12%	24%

[1]Data from 1995 on are estimates or projections.
Source: Population Projections of the United States by Age, Sex, Race, and Hispanic Origin: 1995 to 2050 (U.S. Bureau of the Census, 1996).

Figure 3-5

Travel Behavior of Men 16 and Older by Age, 1990

Age Category	Average Daily Person ...		Average Daily Vehicle ...	
	Trips	-Miles	Trips	-Miles
16–19	3.48	29.77	1.85	15.56
20–29	3.54	39.68	2.76	29.77
30–39	3.37	39.53	2.88	31.91
40–49	3.15	40.42	2.76	31.76
50–59	2.91	35.05	2.51	26.99
60–64	2.87	26.85	2.49	22.44
65+	2.24	18.13	1.84	14.07

Source: 1990 NPTS: Demographic Special Reports (Federal Highway Administration, 1995).

baby boom's impact on travel will be waning as 2010 approaches. Boomers made up 35 percent of the population in 1990; by 2010 they will account for only 25 percent. Also, they will be starting to retire, which will mean fewer commuting trips. The NPTS results clearly indicate that travel activity declines after age 50. If the travel habits of current retirees carry over to the baby boom generation, growth in total travel should begin to slow before 2010.

However, there is no certainty that baby boomers will follow the travel patterns of the previous generation. Boomers are likely to be more active, and they are certainly more accustomed to frequent driving than were their predecessors.

The elderly population will be growing more rapidly than the population in prime driving age groups. In the aggregate, an aging population will mean reduced per capita travel demand. Women live longer and thus outnumber men in the senior population. Because women tend to make fewer and shorter trips than men, an aging population suggests a drop in total travel.

However, the differences in travel demand between younger people and seniors will be narrowing. According to the 1990 NPTS, the biggest percentage growth in travel mileage for all age groups was registered for persons in the over-70 group (up 40 percent on average). Seniors will continue to take fewer trips per capita than the national average. However, their travel patterns are changing. From 1983 to 1990, seniors (persons 65 or over) increased their total travel mileage on average by 20 percent, even though they made only 6 percent more trips per capita. While seniors make fewer work and shopping trips, pleasure travel (usually for longer distances) increases

after retirement. Recreation- and leisure-related trips will grow in importance as active baby boomers leave the full-time work force.

The image of seniors traveling by bus or on foot to take care of their daily needs is no longer accurate. Older people now make at least three-fourths of their trips in private vehicles, either as drivers or passengers. As a group, urban area seniors in 1990 made only 2 percent of their trips using public transportation. Walking is a far more common means of travel, but even this is declining dramatically. For example, persons aged 80 to 84 made only 13.6 percent of their trips on foot in 1990, compared with over 22 percent in 1983.

More seniors will be driving well into their retirement years. A significant percentage of women born before 1930 never learned to drive. By 2010, 90 percent of women aged 65 and over (and nearly 100 percent of men) will hold driver's licenses. They will be reluctant to give up their cars as they age—and for good reason. The elderly population by and large is no longer living in densely populated urban neighborhoods with convenient access to public transportation and a concentration of activities. Until 1980, a majority of urban area seniors lived in such neighborhoods. Most now live in suburbs where opportunities for shopping, personal services, and socializing may not be within walking distance of their homes. The old neighborhood way of life will become even less common for seniors as suburban baby boomers begin to retire in large numbers after 2010.

More people living beyond age 75 will result in more trips taken by younger family members to help with their care. In 1980, 4.4 percent of Americans were age 75 and older; by 1995, this age group's share had risen to 5.6 percent and it is expected to reach 6.2 percent by 2010. Seniors tend to want to remain in their own homes as long as they are physically able to do so. People travel far less as they become more frail, but their younger relatives, neighbors, and friends travel more to assist them. Demand for paratransit services from transit agencies, taxi companies, or social service providers will expand dramatically, especially in suburban communities that lack neighborhood shopping and services.

Gender, Ethnic, Lifestyle, and Income Factors

Aside from the aging of the population, other demographic characteristics also influence travel behavior. Trip rates, travel mileage, and commuting preferences vary by sex, race, household composition, and income.

The 1990 NPTS and other research suggest continued divergence in travel behavior between men and women. Although men constitute less than half of the population, they account for more than half of all trips and mileage. However, the differences in travel behavior are narrowing. The proportion of women with driver's licenses is approaching that of men. Women are traveling greater distances, making more trips, and doing more of their travel by car than ever before. Between 1983 and 1990, personal vehicle trips for both men and women increased by about 25 percent, while total trips for women increased by 14.9 percent compared with only 7.3 percent for men. Men are still averaging more miles on the road: vehicle-miles traveled by men jumped nearly 50 percent in the seven-year period, compared with an increase of just 28.5 percent for women.

It appears that men are commuting ever increasing distances to and from their jobs, while women are working closer to home. At the same time, though, women are making more trips for shopping, childcare, and family or personal purposes. (See Figure 3-6.) Women are also more likely than men to practice trip "chaining," which is visiting multiple destinations during a single trip. The dedensification of development, the dispersal of shopping concentrations, and the failure to provide pedestrian accessibility at activity centers are likely to result in further increases in vehicle trips by women.

Married men with children, regardless of the age of their youngest child, make the same number of trips as married men without children.[6] Married women with children make more trips than their spouses do, according to the 1990 NPTS. The category of households with children has been declining as a share of all households for years, but its share will stabilize in the near future. Single, working mothers generally travel farther and use a car more than married mothers. The percentage of all households consisting of single mothers with children rose from 4.5 percent in 1990 to 5 percent in 1993. Growth in female headed households is likely to slow, but families headed by single fathers will be a growing share. The travel patterns of custodial fathers will resemble those of single mothers more than those of married fathers.

The use of transit and carpooling is declining, more so for women than for men. Studies suggest that women with children need the flexibility that driving their own car to work affords them. Driving allows them to drop off or pick up children at school or day care and to conduct other errands on the trip home. For women without children, declining use of transit may reflect the movement of clerical jobs out of central business districts into outlying city neigh-

Figure 3-6
Trip Purposes by Gender, 1990[1]

Trip Purpose	Percent of All Person Trips[2]		Percent of Total Vehicle-Miles Traveled	
	Male	Female	Male	Female
To/From Work	24.1%	18.0%	39.7%	32.2%
Work Related	1.6	0.9	2.5	1.2
Shopping	17.0	21.5	9.9	14.7
Other Family/ Personal	19.6	23.8	17.3	23.4
Civic/Educational/ Religious	10.9	11.3	4.0	5.4
Vacation/Pleasure	0.4	0.3	1.0	0.4
Visit Friends/ Relatives	9.5	10.0	10.1	11.8
Other Social/ Recreational	16.2	13.7	14.8	10.3
Other	0.7	0.6	0.7	0.6
Total[3]	100.0	100.0	100.0	100.0

[1]Data are for urban area households.
[2]Includes all modes of travel.
[3]Totals may not add due to rounding.
Source: 1990 NPTS: Urban Travel Patterns (Federal Highway Administration, 1994).

borhoods and suburbs, where transit service is less frequent or not even available.

Race and ethnicity further differentiate travel patterns. Non-Hispanic white men and women make the most trips and travel the longest distances, on average. (See Figure 3-7.) Whites are a shrinking percentage of the total U.S. population. Hispanics are expected to increase their share of the population from 9 percent in 1990 to 13.8 percent by 2010.

As whites become a smaller proportion of the population, overall per capita travel might be expected to decline. But the impact of changing ethnic composition on travel is not so simple to predict. It is important to note that travel by Hispanic males is growing faster than travel by other racial/ethnic groups. Average daily vehicle trips by Hispanic males more than doubled between 1983 and 2000, and their average daily vehicle-miles traveled increased by 156 percent. On the other hand, Hispanic women are far less likely to hold driver's licenses than are women from other racial/ethnic groups, and they travel less frequently than Hispanic men.

Wealthier households make more trips, use cars more, and travel longer distances. According to the NPTS, households earning under $15,000 in 1990 averaged fewer than five trips per weekday.

Figure 3-7

Travel Behavior of Persons 16 to 64 by Race and Sex, 1990

	Average Daily Person ...		Average Daily Vehicle ...	
	Trips	**-Miles**	**Trips**	**-Miles**
Hispanic				
Men	2.8	29.5	2.0	18.9
Women	2.7	17.4	1.4	9.0
White				
Men	3.4	38.9	2.8	27.8
Women	3.7	31.1	2.6	17.8
Black				
Men	3.0	24.1	2.0	16.1
Women	3.1	19.7	1.8	11.0
Other				
Men	2.8	23.5	1.9	16.9
Women	2.8	16.8	1.5	9.1

Source: 1990 NPTS: Demographic Special Reports (Federal Highway Administration, 1995).

Those with incomes over $50,000 averaged well above 9.5 trips. Not surprisingly, low- and moderate-income households made fewer vehicle trips. They were also more likely to have more than one person in the car per trip. Men with higher household incomes not only made more vehicle trips and traveled greater distances, but also they registered the largest increases in these indicators of travel activity between 1983 and 1990.

These disparities will not change. Families with limited incomes have fewer vehicles at their disposal and cannot afford to drive them as frequently as families with high incomes. However, the differences in travel behavior between the poor and the wealthy may narrow in the future. The continued migration of jobs to outer-ring suburbs (see later sections) means that low-income city residents will have to travel longer distances to stay employed. The further exodus of shopping and healthcare services from inner-city neighborhoods would also necessitate longer trips.

For middle- and upper-income households, a greater use of electronic communications technology is likely to reduce the demand for business trips as well as routine personal travel. Videoconferencing, while not currently widespread, could eventually be substituted for many on-site meetings and business contacts requiring travel.

Regional Population Shifts

The movement of population and households to the Sunbelt continues, although at a somewhat slower pace than in the past. All regions of the country will grow over the next 15 years, but by 2010 the Frostbelt will hold less than 40 percent of the total U.S. population. (See Figure 3-8.) (In 1970, more than half of all Americans lived in the Northeast and Midwest).

Between 1990 and 1994, the population of the western states increased 7.7 percent, and, for the first time, most of this growth occurred outside California.[7] The Mountain states are now experiencing the impact of rapid population growth on road systems that have been slow to catch up. In 1993 alone, the number of people in Nevada expanded by 5.4 percent. The six fastest-growing states in the early 1990s are in the West. Movement to the southern states is also continuing. The South, which includes Texas and Florida, grew by 6.1 percent between 1990 and 1994.

Americans are moving from more densely developed regions to areas where housing and employment centers are more dispersed. Density in the Mountain states averaged only 17 persons per square mile in 1993, compared with 283 persons per square mile in the Middle Atlantic states. This has important implications for the feasibility of mass transit, and the ability to maintain the share of total trips taken on foot or by bicycle. The 1990 Census indicated that 12.8 percent of commuters in the Northeast used transit, compared with only 2.6 percent in the South and 4.1 percent in the West. The national average of

Figure 3-8

U.S. Population by Region, 1960–2010[1]

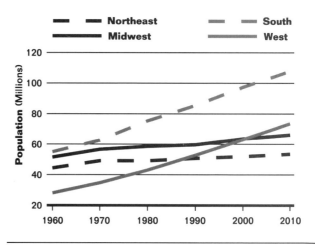

[1]Data from 1995 on are projections.
Source: Current Population Reports P25-1111 (U.S. Bureau of the Census).

commuters using transit was only 5.3 percent. Even more important for transportation planners, low-density development also means longer trips to the store, to school, and to work.

The Move to the Suburbs

During the 1980s, the population of central cities grew at a rate only one-third that of the suburbs. Put another way, about three-fourths of metropolitan area growth occurred in the suburbs. By 1990, the suburban share of total U.S. population stood at 47 percent, up from 43 percent in 1980. As seen in Figure 3-9, this trend to the suburbs has been in evidence since the 1950s.

The rate of suburbanization is slowing, but not because people have decided to stay in central cities or move back from the suburbs. The general slow-down in population growth and new household formations is largely responsible, as well as the settlement of large numbers of immigrants in central cities during the 1980s. Some central cities, such as New York and Los Angeles, have stable populations because of immigration. For the foreseeable future, however, not many of these cities will be able to halt the exodus of jobs and households.

Low-density development means more automobile trips. Outward movement from densely populated central cities and mature suburbs to newly developing communities at the urban fringe explains much of the increase in private automobile use documented in the NPTS. More than 87 percent of the 240 billion person trips made in 1990 were in private vehicles, up from less than 82 percent in 1983. Walking was the second most popular travel mode among urban area residents, accounting for 7.4 percent of person trips; but this represented a drop from its 8.6 percent share in 1983. Bus and rail trips fell from 2.3 percent of all trips in 1983 to just 1.8 percent in 1990.

Although the growing use of private automobiles can be seen as an expression of the desire of Americans for privacy and convenience, the fact is that many suburban areas offer no alternatives to driving. Even in the suburbs of large metropolitan areas with well-established commuter-rail and bus systems, transit is an option only for the journey to work. For other daily activities, such as shopping or doctor visits, suburbanites drive because they have to. Housing has been deliberately separated from commercial nodes by suburban zoning, and many residential areas lack sidewalks or other means of pedestrian access to shopping and services that may be close enough to reach on foot.

Nonwork travel will increase as a share of total trips. Commuting in 1990 accounted for only one-

Figure 3-9

The Changing Geographic Location of U.S. Population, 1950–1990

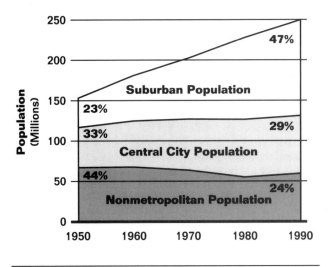

Source: Commuting in America (Eno Transportation Foundation, 1996).

fourth of total trips. One of the most important findings from the 1990 NPTS is that family and personal needs have become the biggest generator of trips, overtaking jobs. Preliminary data from the 1995 NPTS indicate that shopping trips have become the second most common type of trip, moving work trips down to third place. While commuting trips are longer on average, trips for personal needs other than shopping are now more numerous and are getting longer.[8] Because activities and services are no longer concentrated in neighborhood business districts or town centers, more frequent trips to distant locations are necessary.

The increased number of nonwork trips—coupled with more flexible working hours, more reverse commuting, and other changes in commuting patterns—means that traffic congestion is no longer limited to the peak commuting hours. Traffic management solutions that are more complex than those designed merely to move rush hour workers are now required.

Dispersed development means more shopping trips, while increased catalog and electronic shopping will mitigate against this trend. Shopping trips accounted for just over 19 percent of all trips in 1990 and 13 percent of vehicle-miles traveled by urban area households. The length of the average shopping trip —4.9 miles—was unchanged from 1983 to 1990, but shopping trips slightly increased their share of total trips. Trip chaining appears to be on the rise, as more

Commuting Patterns in Chicago

Within the six-county Chicago region, Cook County (including the city of Chicago) is still the dominant destination for commuting trips. However, the number of persons who live in Cook County but work outside its borders has more than doubled since 1970, while the number of workers who both live and work in the county has grown by only 2 percent. There were significant changes in the journey to work from 1970 to 1990. Of particular note is the increasing share of workers who commute across county borders, the declining share of workers in the city of Chicago, and the emergence of DuPage County as a net importer of workers.

During the 1980s, the region's population grew by only 2 percent, yet commuting trips grew by nearly 13 percent. Explosive growth in the I-88 (East-West Tollway) corridor transformed DuPage from a bedroom community to a business center. DuPage now draws in more workers than it sends to other counties, despite the fact that it is the smallest of the six counties in the area. Kane, McHenry, and Will counties also experienced growth in both imported and exported work trips.

Nearly 563,000 persons worked in private sector jobs located in or near Chicago's central business district (including downtown, near north, and near south side areas) in 1990.[1] The number of private sector jobs in the central area grew by over 7 percent between 1985 and 1990, and job growth in the downtown core was nearly 10 percent. According to the 1990 Census, 29 percent of workers who live in the city worked downtown, 47.5 percent worked elsewhere in the city, and only 23.5 percent commuted to the suburbs. In contrast, only 9.9 percent of suburban jobholders worked downtown. The overwhelming majority of suburban workers (78.8 percent) held jobs outside the city.[2]

The recession that began in 1989 brought some different employment growth dynamics into play. The city of Chicago as a whole began to lose jobs, although the downtown job market continued to grow until 1990. By 1993, downtown had lost nearly 63,000 jobs. The rate of job loss downtown and elsewhere in the city exceeded that of the suburbs. Both the city and the CBD registered modest employment gains in 1994, but neither has returned to prerecession peaks. Meanwhile, the number of jobs in the region's five edge cities continues to grow.[3]

Chicago's downtown remains the hub of the regional transportation system. Frequent bus and subway service is provided from city neighborhoods, and the suburban commuter-rail system is extensive. Despite this, the number of work trips via transit regionwide dropped 10 percent during the 1980s. Much of the loss can be attributed to the increase in commuting between suburbs, where transit service is less convenient. Again, DuPage County is illustrative. In 1980, 10 percent of commuters living in DuPage used mass transit, mainly commuter-rail service to downtown. By 1990, this share had dropped to 7 percent. During the same period, total work trips originating in the county grew by 27 percent.[4]

Notes

1. Based on data from the Illinois Department of Employment Security for ten Chicago zip codes bounded by North Avenue on the north, Halsted or Ashland on the west, 35th Street on the south, and Lake Michigan on the east. In the core business district (the seven zip codes north of Van Buren), 1990 private sector employment was estimated at 487,125, up 9.7 percent from 444,196 in 1985. See Illinois Department of Employment Security, *Where Workers Work in the Chicago Metropolitan Area* (annual).

2. Patrick T. Reardon, "Loop Still Has Pull Even as City Jobs Slip," *Chicago Tribune*, April 12, 1995, p. 1.

3. These are commonly known as the Edens Expressway Corridor and the O'Hare Airport area (Cook County); the Schaumburg/I-90 corridor (in both Cook and DuPage counties); the Tri-State Tollway node (primarily in Lake County); and the I-88 corridor (DuPage and Kane counties).

4. See "Change in Travel Mode to Work from 1980 to 1990," *Transportation Facts* (Chicago Area Transportation Study), v. 10, no. 6, October 1993.

---

auto commuters combine the journey home from work with shopping stops or as shoppers visit more than one destination in a single outing.

Growth in shopping travel is likely to continue, but catalog shopping and electronic shopping will offset some of the increase that would otherwise occur. Although such shopping alternatives now account for only a small share of total retail sales,[9] their share will grow as the technology becomes more sophisticated and customer service is improved. On the other hand, the recent explosion of big-box retailing and power centers—much of which occurred after the 1990 NPTS—may well put many neighborhood retailers out of business and thus cause shoppers to travel longer distances to stores.

Cities are capturing a decreasing share of new jobs, resulting in longer work trips and lower public transit ridership. During the 1980s, jobs opportunities expanded dramatically in both central cities and suburbs. The long-term trend, however, is a steady increase in the share of total employment located outside of core cities. In 1980, 38.1 percent of workers commuted to jobs in the central city of the metropolitan area in which they resided. By 1990, the share had dropped slightly to 37.6 percent. Of the 18.5 million persons added to the work force during the decade, 7.2 million (39 percent) found work in suburbs, while 6.5 million (35 percent) found work in central cities. The remainder found work in nearby metropolitan regions or outside metropolitan area boundaries.[10]

In the nation's largest metropolitan areas, the suburbanization of jobs has been most pronounced. In the 35 metropolitan areas with populations of more than 1 million in 1983, the share of workers commuting to jobs in their central city has steadily fallen: from 48.4 percent in 1970, to 43.0 percent in 1980, to 38.3 percent in 1990.[11] Since 1960, suburbs have captured nearly all of the growth in population and jobs in these 35 regions. Five of every six jobs added were located beyond the city limits.

During the 1990s, cities will not be as successful as they were in the 1980s in slowing the rate of exodus of jobs to the suburbs. In the 1980s, the rate of population loss from cities declined and urban centers experienced more employment growth than in the previous two decades combined. In many cities, the dramatic expansion of downtown office employment more than offset the disappearance of inner-city factory jobs.

But suburbs are the future. Rapid changes in communication technology not only make it possible for people to work at home, but also make it unnecessary for many firms to be downtown. Business people can handle routine communications and obtain information electronically from remote (and less costly) locations. Also, lifestyle choices are increasingly important in determining where workers will look for job opportunities. Today's time-conscious worker wants to spend less time on the road or in the bus.

Downtowns were the premier office location in the 1970s because they had a virtual monopoly on top quality space with on-site and off-site amenities. Downtown buildings maintained lower vacancies than their suburban competitors despite higher rents and operating costs. But during the real estate boom of the 1980s, the quality of suburban space improved. Edge cities emerged with Class A office accommodations, good restaurants, hotels, health clubs, nearby shopping, and cultural events. These employment centers were surrounded by the well-educated, white-collar work force that employers sought, and they usually enjoyed good highway access. Executives began to ask why they should subject themselves—and their employees—to long daily journeys downtown.

Suburban office properties came out of the real estate recession at the same time that office tenants realized that they no longer really needed to be downtown. At the beginning of 1994, nationwide suburban office vacancy rates dipped below those for downtowns for the first time. This trend continued through the first half of 1995. Even though downtown office construction remains moribund, vacancy rates were still above 20 percent in many CBDs at the end of the second quarter 1995—including Baltimore, Dallas, Detroit, Hartford, Houston, San Diego, and Tampa.[12]

The shift of jobs to edge cities means quicker commutes for suburbanites, more travel for lower-income city residents, and lower transit demand. As with other demographic trends affecting travel,

the impact of the suburbanization of the workplace is not uniform. Although between 1983 and 1990 commuting *distances* went up 22.4 percent—from 8.3 miles to 10.1 miles—average commuting *time* increased only 6 percent. To the extent that edge cities provide job opportunities closer to where new housing is being built, workers who live nearby can shorten both their commuting times and distances, in part because driving at the suburban fringe is faster than riding mass transit for an equivalent distance. To the extent that these people are shifting from transit to cars, their commuting times will tend to decrease. The opposite is true, however, for workers who reside in the inner city. They have to travel further, and few convenient reverse commuting transit options are offered.

The dispersal of employment centers means more commuting between suburbs and across county lines. Commuting between suburban locations and between counties is increasing. The decennial Census provides the most detailed information available on where people work in relation to where they live. Figure 3-10 presents data from 1980 and 1990 for working residents of counties in four metropolitan areas—Austin, Charlotte, Sacramento, and Pittsburgh. The first three are rapidly growing regions with well-educated work forces and a high percentage of employment in high-tech industries. Two are state capitals as well. Greater Pittsburgh was the only one of the nation's 35 largest metropolitan areas to lose jobs during the 1980s. A number of patterns are apparent. In all but one county (in the Pittsburgh area), commuting across county lines is a growing phenomenon. In most counties, the share of workers who do so is large; only for residents of the core counties is cross-county commuting uncommon. The Austin area experienced the biggest increase in cross-county commuting and shows the highest rates of movement among the suburban counties. Pittsburgh showed the least change.

Labor Force and Workplace Changes

Even though commuting trips represent a decreasing share of total travel, the longest and most time-consuming trips (outside of vacation travel) involve the journey to work. Labor force growth trends do not precisely follow population patterns. As with population growth, the labor force is expanding, but at a slower rate than in previous decades. A slowdown in employment growth should mean less pressure on overburdened road systems, but, as indicated above, job growth is likely to be concentrated in suburban areas with less capacity to handle it.

Figure 3-10

Cross-County Commuting in Selected Metropolitan Areas, 1980 and 1990

MSA/County[1]	Workers Commuting Out of Their County Of Residence (Percent)	
	1980	1990
Austin, Texas		
Bastrop	43.9%	54.7%
Caldwell	38.1	48.8
Hays	30.2	45.9
Travis*	3.5	5.0
Williamson	54.1	60.8
Charlotte, North Carolina		
Cabarrus	30.5	42.9
Gaston	23.1	26.1
Lincoln	40.2	48.2
Mecklenburg*	5.7	6.6
Rowan	27.9	30.5
Union	36.3	39.3
York (South Carolina)	25.6	35.3
Pittsburgh, Pennsylvania		
Allegheny*	5.5	6.7
Beaver	28.9	39.0
Butler	26.9	32.0
Fayette	34.1	32.0
Washington	32.4	36.5
Westmoreland	33.5	36.1
Sacramento, California		
El Dorado	37.2	42.3
Placer	38.2	42.8
Sacramento*	7.7	11.9
Yolo	28.5	33.2

[1]Counties in MSA as defined in 1994.
*Represents county in which the central city is located.
Sources: 1980 Census; and 1990 Census.

Growth of the labor force is slowing. During the 1980s, over 18 million persons entered the labor force, which grew by nearly 20 percent. In contrast, the overall population grew less than 10 percent. Labor force growth was fueled during the early part of the decade by the tail end of the baby boom reaching working age. Also, the labor force participation rate of females increased from 51.5 percent to 57.5 percent over the decade.

A number of factors account for the impending slowdown in the growth of the labor force. Most im-

portantly, the huge baby boom cohort has passed the age of entry into the work force, and the next generation of workers is much smaller in size. Growth in the female labor force is leveling off, and male labor force participation rates are actually declining. In 1994, 4.6 million men between the ages of 25 and 54—the prime work years—were not in the labor force. The Bureau of Labor Statistics (BLS) indicates that only 75 percent of adult males were in the labor force in 1994, compared with 80 percent in 1970. In the not too distant future, participation rates for men and women could converge, with women constituting nearly half of all workers. As mentioned earlier, commuting trips by women tend to be shorter than those by men.

More people are holding two or more jobs. Corporate downsizing and outsourcing, wage reductions, and other workplace trends have resulted in more people holding multiple jobs. According to BLS, 7.9 million workers held two or more jobs in June 1995. Although this represents under 7 percent of all workers, the numbers are growing and are probably understated in official statistics.[13] People working two or more jobs tend to have irregular schedules that make carpooling difficult. Moonlighting results in greater auto use, but not necessarily during peak hours.

More companies are offering flexible work hours as an employee perk and to reduce peak hour traffic. In 1991, over 15 percent of full-time workers reported that they could take advantage of flextime options at their place of employment, up from 12.3 percent in 1985. As with part-time work, flexible hours help avoid rush hour gridlock. However, they also make carpooling less feasible and transit use less convenient, in that buses and trains do not run as often in the predawn or early evening hours.

Telecommuting and working at home are increasingly popular options. Another way that employers can help reduce rush hour traffic (and shrink office space needs at the same time) is to offer telecommuting or other work-at-home options. At present, very few employees telecommute full time, although telecommuting has received considerable attention in the press. A 1991 U.S. Census Bureau survey indicated that nearly 20 million people did some work at home, but only 5 percent did it full time, either as salaried workers or in their own home-based businesses. (See Figure 3-11.) Most of the home workers discussed in the media are people who bring home extra work after normal office hours or who run part-time businesses from their homes. Nevertheless, both full- and part-time telecommuting are on the rise. One estimate suggests as many as 9 mil-

Figure 3-11

Persons Working at Home, 1991

	Number	Percent
All Home Workers[1]	19,967,000	100.0%
Paid for Work at Home	7,432,000	37.2
Paid for Full-Time Work	1,070,000	5.4
Self-Employed	976,000	4.9
Wage and Salary Workers	94,000	0.5
Paid for Part-Time Work	6,362,000	31.9
Not Paid for Work at Home[2]	12,165,000	60.9

[1]Total exceeds sum of reported categories because some workers did not report their pay status.
[2]Wage and salary workers who bring home extra work.
Source: Monthly Labor Review, February 1994.

lion workers telecommuted for at least part of their workweek in 1994, and some researchers expect telecommuting to experience double-digit annual percentage increases for the balance of the decade.[14]

Of all of the workplace changes now occurring, telecommuting offers the greatest potential for reducing total travel and is the only change that is likely to cut automobile use. While telecommuting is gaining widespread acceptance in the communications industry,[15] most employers have yet to be sold on the idea, and many workers are not happy with it. Employees miss interaction with colleagues and find that being visible is important to their upward mobility. Home-office work is likely to grow steadily, but not at the rates predicted by the office equipment industry. It is not a panacea for reducing work-related travel.

Employers seeking compliance with federal Clean Air Act mandates pioneered flexible work hours and work-at-home options. Such initiatives will reduce peak hour travel in the short run, but they also lengthen the duration of the rush hour and may increase total trips. In the long run, less traffic during rush hours could entice people who work during traditional hours away from mass transit, vanpools, and carpools and back into private autos. With Clean Air Act enforcement either inconsistent or nonexistent in most states, it remains to be seen if the flexible workplace will actually come into being in a big way and reshape commuting patterns.

A Mixed Picture

Demographic trends indicate that traffic volume will continue to grow, although not at the rates experienced during the last three decades. A number of

factors will work to slow the increase in total trips and peak hour congestion:

- An aging population;
- Slower growth in the number of drivers and vehicles;
- More trip chaining;
- More shopping via catalog, telephone, and computer;
- Stabilized female labor force participation rates;
- Slower employment growth; and
- Telecommuting, home offices, and flexible work hours.

Work-related trips will continue to decline as a share of total travel. However, other trends will push travel demand up:

- Thirty-five million more Americans (by 2010), whose travel needs must be accommodated;
- Immigration concentrated in the prime driving age groups;
- More travel by women, especially those with young children;
- More leisure travel by senior citizens, and a higher percentage of retired women with driver's licenses; and
- Longer shopping trips, as power centers and outlet malls serving large trade areas compete with neighborhood retailing.

Perhaps most important for transportation planning is the shifting location of population and jobs. Population movement from the Northeast and Midwest to the South and West, employment growth in edge cities, housing development in areas remote from shopping and services—all suggest more trips, greater reliance on private automobiles, fewer trips made on foot, and less transit use.

Notes

1. The Federal Highway Administration has published a series of reports highlighting findings from the 1990 Nationwide Personal Transportation Survey and changes in travel behavior. The findings reported in this chapter are from *1990 NPTS: Urban Travel Patterns*, FHWA PL-94-018, June 1994, and *1990 NPTS: Demographic Special Reports*, FHWA PL-94-020, February 1995 (Washington, D.C.: U.S. Department of Transportation).

2. The U.S. Census Bureau's mid-range projection series assumes that net immigration will remain level at 820,000 persons per year. Immigrants as a group have fewer children and elderly persons, suggesting greater than average travel demand. However, it can take time for newcomers to get a driver's license and an automobile. See table C on page 5 of *Population Projections of the United States by Age, Sex, Race, and Hispanic Origin: 1995 to 2050*, Current Population Reports P25-1130 (Washington, D.C.: U.S.

Bureau of the Census, 1996) for a summary of the mid-range projections.

3. Joel E. Rey et al., "An Assessment of the Potential Saturation in Men's Travel," in *1990 NPTS: Demographic Special Reports*, p. 1-15.

4. Given that some senior citizens have never learned to drive or chose not to renew their licenses when they stopped driving, it is clear that virtually all Americans who want to drive are already doing so.

5. U. S. Bureau of the Census, *Statistical Abstract of the United States: 1995* (Washington, D.C.: U.S. Government Printing Office, 1995), p. 634.

6. See Sandra Rosenbloom, "Travel by Women," in *1990 NPTS: Demographic Special Reports*, p. 2-16. See also Steven W. Rawlings, *Household and Family Characteristics: March 1993*, Current Population Reports P20-477 (Washington, D.C.: U.S. Bureau of the Census, 1994) for historical statistics on household size and composition.

7. California's growth rate in the 1990s has been less than the national average; during the 1980s, the population of California grew at twice the rate for the United States as a whole.

8. The finding that trips for family and personal needs outnumber commuting trips is based on both weekday and weekend trips under 75 miles by urban area residents. For weekday travel, the journey to work remains the most common purpose for travel.

9. In 1994, mail-order sales totaled over $46 billion, which was only 2.7 percent of total nonautomotive retail sales.

10. See Alan E. Pisarski, *Commuting in America*, 2nd ed. (Washington, D.C.: Eno Transportation Foundation, 1996) for further analysis of changes in work location since 1980.

11. See Richard L. Forstall, "Commuting Trends in Major U.S. Metropolitan Areas: 1960–1990" (paper presented to the Southern Demographic Association, October 30, 1992). Percentages are based on metropolitan area definitions as of 1984, which did not change prior to 1990. Forstall, a U.S. Census Bureau staff member, notes that in 1960 the majority (56.4 percent) of workers residing in these 35 areas still worked in central cities.

12. See CB Commercial's *Office Vacancy Index of the United States*, June 30, 1995. Downtown vacancy rates exceeded suburban rates by five percentage points or more in Albuquerque, Atlanta, Austin, Baltimore, Cleveland, Dallas, Detroit, Fresno, Honolulu, Indianapolis, Jacksonville, Kansas City, Nashville, Oklahoma City, Phoenix, St. Louis, Tampa, Tucson, and Wilmington. Downtown vacancy rates were lower than those of the suburbs in only 12 metropolitan areas covered by the CB Commercial survey.

13. According to a 1991 Gallup poll, one in six full-time workers held second jobs. Given that 1991 was a recession year, the percentage in 1996 is undoubtedly higher.

14. According to press reports, the Emerging Technologies Research Group of Ithaca, New York, estimates that there were 9.2 million telecommuters in 1994, up from 7.6 million in 1993; and Link Resources of New York City estimates that there were 6.6 million telecommuters in 1994, up from just 2.4 million in 1990. Counting consultants and independent contractors who telecommute would raise such estimates by 25 to 30 percent.

15. At AT&T, for example, 35,000 employees (nearly 20 percent of the work force) are said to be telecommuting one or more days each week. See Barbara Presley Noble, "Nudging Workers from Comfy Nests: AT&T Pushes for Telecommuting," *New York Times*, July 30, 1995, sec. 3, p. 10.

Chapter 4

Transportation and Development in Portland
Challenging the Idea of Laissez-Faire Development

You have a basis here for civilization on its highest scale. . . . Have you got enough intelligence, imagination, and cooperation among you to make the best use of these opportunities?

—Lewis Mumford, Portland City Club speech, 1938

Portland, Oregon, has been praised as "a metropolitan area that works, a model for other urban regions seeking a regional solution to difficult growth issues. Its downtown and inner-city neighborhoods are robustly alive. Development is controlled within an urban growth boundary, preserving farmlands forests, and access to natural resources. The transportation system has shifted determinedly from dependence on highways to a balance of alternatives. Workable affordable housing programs are in place. And the region operates under an effective metropolitan governance system that is unique in America."[1]

Among the most notable initiatives in Portland in the areas of land use, growth management, and transportation are the setting of a lid on downtown parking (1972), the establishment of a regional government whose members are elected (1978), the adoption of an urban growth boundary (1980), the construction of a light-rail line (service beginning in 1986), and, at the state level, the adoption of the Transportation Planning Rule, a requirement for transportation planning at the metropolitan level (1991).

These policies would be considered quite radical in other communities. In combination, they signal the community's willingness to buck current trends and challenge the idea of laissez-faire development in order to make a better life for all. Quality of life is an amorphous concept that may be difficult to measure, but it is clear to residents and visitors alike that Portlanders enjoy a very high quality of life. Aggressive regulations have been responsible at least partly for this.

The region's support for transit has resulted in a popular and widely praised transit system, built around the Metropolitan Area Express (MAX) light-rail line. Combined bus and rail transit ridership increased 26 percent during the 1980s, although the number of commuters using transit as their usual mode of getting to work declined by 18 percent. Car use has grown faster than transit use, with vehicle-miles driven having increased almost five times as fast as population between 1980 and 1990. While the Portland region offers one of the better U.S. metropolitan area examples of organizing growth to avoid gridlock, it also faces continuing challenges in the transportation area, confirming the complex nature of regional transportation planning and development.

Before the 1970s: Setting the Stage

Concern about uncontrolled suburban development in Portland prompted the naming of a state commission to consider transportation, and other issues. The commission's report suggested that much of the problem was lack of coordination between the city and the county. This is an ordinary enough story today. Many growing communities face the same problem and have reached the same conclusion. The difference is that the Portland commission made its findings in the 1920s.[2] Planning and growth management is a mature industry in Portland, and indeed in Oregon.

Downtown Focus

After World War II, businesses started locating in the suburbs.[3] The Portland Development Commission

Figure 4-1
Population of Portland, 1970–1995

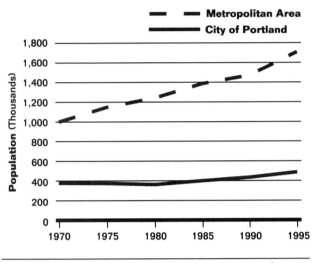

Sources: Decennial Census; and Portland State University, Center of Population Research and Census.

was established in 1958 to manage an urban renewal initiative, the South Auditorium Project.[4] Portland's early attempts at downtown revitalization scored some successes, but planners also made the kinds of mistakes that cities across the country were making. Slum clearance robbed downtown of much of its character, architectural history, and housing. Neighborhoods were rendered unstable. Demolishing historic structures created a large number of surface parking lots that isolated remaining buildings and gave the city a gap-toothed look.[5] Downtown became increasingly vacant, polluted, and less desirable as a place to visit and conduct business. George Sheldon, chairman of the city planning commission in the 1970s, decried the fact that downtown had "more parking lots than buildings."[6] However, the city studied its mistakes and made new plans and became, in one observer's view, the Lazarus of American cities.[7]

Regionalism's Beginnings

In response to "wasteful, fragmented, and uneven urban services" and a "fragmented local government," a group called Metropolitan Area Perspectives (MAP) came into existence in January 1961.[8] Its promoters were citizens, planning professionals, and business-persons. MAP was the first organization to address the idea of establishing a regional government for the Portland area. Its efforts led to the creation of the Portland Metropolitan Study Commission (PMSC), which functioned from 1963 to 1971. Consisting of

38 legislators representing Multnomah, Clackamas, Washington, and Columbia counties, the PMSC was primarily responsible for preparing a comprehensive plan for furnishing services in the Portland metropolitan area. Through its efforts, several important regional agencies were established: the Metropolitan Service District (MSD) in 1970, the Columbia Region Association of Governments (CRAG) in 1966, and the Portland Metropolitan Area Local Government Boundary Commission (the Boundary Commission) in 1969.

Directed by a seven-member board composed of elected officials, the MSD was intended to be a "box" that could hold as many service responsibilities as the state legislature or local voters were willing to put in it. An attempt to pass state legislation creating the MSD had failed in 1967; passage in 1970 involved the strong support of groups like the League of Women Voters, the Home Builders Association, and the Chamber of Commerce. In its early years, the MSD was predominantly involved in solid waste planning and the management of the Washington Park Zoo.[9]

CRAG was structured as a council of governments representing the area's cities and counties, while the governor appointed the members of the Boundary Commission, who could not be elected officials. The Boundary Commission and CRAG set the table for an enormous expansion of regional planning and regional services delivery in the politically turbulent 1970s. (See "Regional Government and the Urban Growth Boundary" in the next section of this chapter.)

Current Conditions and Recent Initiatives: Aggressive Policies and Conventional Markets

Portland Compared with Other Metropolitan Regions

A 1994 comparison of 55 large urban areas pegged Portland as the Ozzie and Harriet of regions—generally in the middle across a wide range of transportation, housing, and regional growth indicators.[10] Many advertisers use Portland for test marketing for that reason.

Portland's reputation for environmental quality is strong. However, auto use in the region is only a little below average and transit is only a little above average among large urban areas (see Chapter 1). In its driving/transit use ratio, Portland is most similar to Cleveland.

As in most other areas with a serious to marginal air quality problem, transit commuting in Portland

Figure 4-2

Transportation Indicators for Portland, 1970–1990

| | 1970 | 1980 | 1990 | Percent Change | |
				1970–1980	1980–1990
Population	1,009,127	1,298,000	1,477,895	29%	14%
Commuters	393,331	568,916	724,532	45	27
In Private Vehicles	327,975	473,797	623,518	44	32
Driving Alone	287,478	418,396	575,492	46	38
Passengers	40,497	74,912	47,576	85	–36
By Transit	22,818	47,484	38,837	108	–18
Commute Share					
Transit	6%	8%	5%	44	–36
Private Vehicles	83%	83%	86%	0	3
Travel					
Daily Vehicle-Miles Traveled (Thousands)	n/a	11,615	19,398	n/a	67
Annual Transit Trips (Thousands)	15,395	44,629	56,066	190	26
Median Single-Family Lot Value	n/a	$22,000	$31,500	n/a	43

Sources: Decennial Census; *Journey-to-Work Trends in the United States and Its Major Metropolitan Areas, 1960–1990* (Federal Highway Administration, PL-94-012); and ULI residential land price surveys.

lost market share between 1980 and 1990, from 8 percent of commuters in 1980 to 5 percent in 1990. In fact, the decline in transit's share of the commuter market in Portland was greater than in Seattle. Transit use declines sharply among persons living farther from downtown. In 1990, 25 percent of the commuters living within two miles of downtown used transit for their trip downtown, compared with only 20 percent of those living seven miles out (see Figure 4-3).

In terms of support for alternatives to driving, population trends indicate some good news and some bad news. As in other cities, a mismatch between where people live, work, shop, and socialize continues to worsen.[11] Between 1980 and 1990, the population living less than one mile from downtown increased by only 1,481, while the number of jobs in the same area increased by 20,580.[12] Downtown contains 20 percent of the region's jobs.[13] About half of the population growth occurring between 1980 and 1990 was within ten miles of downtown Portland. However, areas within three miles of downtown, which represent the best market for transit and probably for walking as well, lost population (see Figure 4-4). Figure 4-5 shows population growth by distance from the light-rail route. Population within one mile of the east rail corridor grew a modest 3 percent, from 199,000 in 1980 to 205,000 in 1990. Slightly fewer people live in the area within one mile of the rail

line's west corridor, but that area grew by a healthier 15 percent. According to the Texas Transportation Institute, traffic congestion in Portland grew steadily throughout the 1980s, to the point where it exceeded congestion in Phoenix by 1992.

In contrast to Portland's policy record on transit and growth management issues, which is one of the most aggressive in the United States, the region's residents are pretty average as travelers, and apparently as housing consumers as well (see "Outlook" section in this chapter). One implication of Portlanders' normalcy is that if solutions to the problems of transportation and development emerge in Portland, they may actually have wide applicability despite the region's reputation as an off-the-wall eco-utopia.

Development Market Trends

Portland has adapted well to changes in the national and international economies. Historically, the region's economy was based on fishing and timber. Expansion more recently into high-tech manufacturing, business services, and finance has meant stable growth for the past ten years and a diversified economy that is less vulnerable to economic cycles.

Manufacturing jobs account for 16 percent of employment. High-tech manufacturing jobs increased from 11,800 in 1970 to 26,200 in 1994. Portland ranks third among U.S. metropolitan areas in high-tech firms per capita, with 1,100 firms including Intel Corporation and Tektronix, Inc. Business services is the largest employment sector, accounting for 26.6 percent of jobs.[14] Employment in this sector has increased 77 percent since 1980.[15] International trade has also become important in the economy. Portland leads the Pacific Northwest in wholesale trade, and is 18th nationally.[16] Overall employment growth in the late 1980s was double the national average, and employment continued to grow during the recession.

The Office Market. Portland's office market is one of the strongest in the country. During the 1980s, Portland added office space at a rate only half the national average, and thus maintained low vacancy rates. In 1993, the office vacancy rate was 11.4 percent, compared with a national average vacancy rate of 17 percent.[17] Valuation Network ranked Portland's suburban office market first among 44 suburban metropolitan office markets in the country, and its CBD office market eighth, based on vacancy rates, estimated years of oversupply, value forecasts, and growth projections.[18]

With 15 million square feet, the central city accounts for 61 percent of the region's office space.[19] In 1993, Class A office space in the CBD enjoyed a low vacancy rate of 8.9 percent.[20] There has been little new construction, so low vacancy rates should continue through 1995. Tenants choose downtown locations for prestige, amenities, and image reasons. Class A office space users downtown tend to be financial and professional firms. Historic office space downtown tends to attract smaller companies in creative fields, such as public relations and architecture.[21] Despite strong competition from the suburbs, the downtown office market should continue to be strong.

Demand for suburban office space can still exceed supply, because large blocks of space are not available. The western suburbs (the Sunset corridor, Beaverton, Washington Square, Kruse Way, Tualatin, and Johns Landing) account for 29 percent of the region's office supply. The eastern suburbs (the east side, the Columbia corridor, Clackamas County, and Sunnyside) account for 6 percent; and Vancouver and Clark County in Washington account for 5 percent.[22]

Figure 4-3

Transit Use by Commuters to Downtown by Location of Residence, 1990[1]

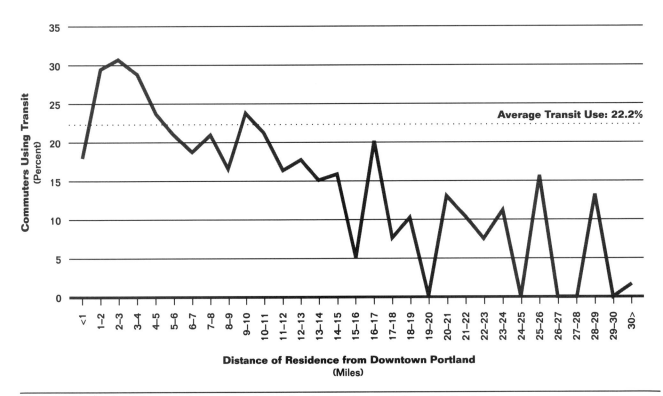

Distance of Residence from Downtown Portland
(Miles)

[1]The universe is workers 16 years and over who work away from home and commute to work in the downtown.
Sources: Metro Planning Department Data Resource Center; and 1990 Census.

Figure 4-4

Population of Portland by Distance from Downtown, 1980 and 1990

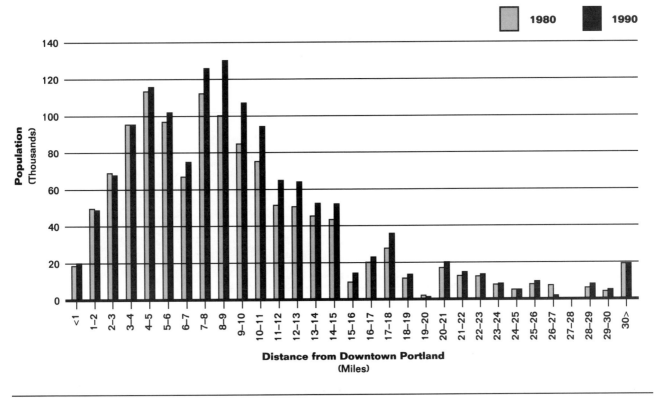

Sources: Metro Planning Department Data Resource Center; 1980 Census; and 1990 Census.

In 1970, according to the Portland Association of Building Owners and Managers, only 10 percent of the region's multitenant office supply was located outside the central city.[23] In the 1980s, the suburban market began offering many more competitive alternatives.[24] According to a 1991 survey of building owners, rents for suburban and central city Class A space were comparable, at between $13 and $20 per square foot including parking for suburban space, and between $12 and $20 per square foot excluding parking for central city space.[25] However, parking is a significant difference: suburban projects offer three to four parking spaces per thousand square feet while parking adds $3 to $4 per square foot to downtown office rents.[26]

The 1991 survey found that suburban office buildings were drawing tenants from sales organizations, manufacturing firms, insurance companies, data processing operations, and residential real estate companies. One reason these types of organizations located in the suburbs was that their sales operations were dependent on auto use and quick access to interstate freeways. The availability of parking was another key reason.

In general, the availability of amenities, considerations of image and prestige, and the cost of the space relative to the building's location are more important factors in development and tenant leasing decisions than is parking. However, when comparable locations are being considered, the availability and cost of parking become critical factors in a leasing decision.[27] Downtown building owners are concerned about the possibility that future office development on sites now providing surface parking could too severely limit parking for their tenants.

The Residential Market. The early 1990s marked the best years for Portland's homebuilders since the 1970s. Most single-family construction occurred near areas in which employment has been increasing, such as Beaverton, Tigard, the Sunset corridor, the Clackamas town center, and the south I-5 corridor.[28] In addition, major projects to rehabilitate pre–World War II housing stock are underway throughout the city. A rising demand for labor should stimulate housing demand over the next five years.

Some people feared that urban growth boundary restrictions on the supply of land for housing would push up housing prices. However, increases in hous-

Figure 4-5

Population of Portland by Distance from a Light-Rail Transit Corridor, 1980 and 1990

| | 1980 | 1990 | Change 1980–1990 | |
			Number	Percent
Total Population	1,241,895	1,412,344	170,449	14%
Population Near LRT East Corridor[1]				
Under 1 Mile	198,574	204,714	6,140	3
1–2 Miles	139,532	146,638	7,106	5
2–3 Miles	69,728	70,666	938	1
3–4 Miles	75,251	77,355	2,104	3
Population Near LRT West Corridor[1]				
Under 1 Mile	138,863	160,044	21,181	15
1–2 Miles	81,200	98,950	17,750	22
2–3 Miles	105,953	112,272	6,319	6
3–4 Miles	130,130	134,852	4,722	4

[1]Some population near downtown Portland is counted in both corridors.
Sources: Metro Planning Department Data Resources Center; 1980 Census; and 1990 Census.

ing prices between 1990 and 1995 have been comparable to national trends, and the region's comparatively low cost of housing contributes to its livability. The median house sales price in 1994 was $109,600, compared with $196,000 in Los Angeles, $177,300 in San Diego, and $152,100 in Seattle.[29] Between 1990 and 1995, the price of a typical residential lot in the Portland region increased by 50 percent, to $50,000, which is a lower rate of land price inflation than experienced in either Houston or St. Louis, for example.[30]

The Retail Market. A growing economy, a growing population, and no state sales tax help make Portland one of the top U.S. metropolitan areas in terms of retail sales per capita.[31] Retail sales grew at a strong annual rate of 7.5 percent between 1985 and 1993. In 1992, total retail sales for the consolidated metropolitan statistical area were close to $14 billion. Multnomah County led with $5.8 billion in retail sales, followed by the city of Portland with $4 billion, Washington County with $3.6 billion, Clackamas County with $2.3 billion, and Clark, Yamhill, and Columbia counties.[32]

National chain stores and category killers are expanding rapidly in the market. The central city retail core, which is oriented to mass transit, is strong, and there are seven regional shopping malls. Retail development is occurring along the future west side light-rail route. The regional retail market contains 29 million square feet of space in 312 regional, community, and neighborhood shopping centers.[33]

In 1990, the Rouse Company and the Portland Development Commission opened Pioneer Place, a three-block retail and office complex. Located on the downtown transit mall, the project takes advantage of front-door service from the MAX light-rail line. Pioneer Place was successful in linking the traditional retail core at the west end of downtown with an emerging speciality niche retailing areas on the east end, restoring downtown Portland as the region's premier retail destination, and catalyzing the revitalization of the surrounding areas.[34]

Strongly growing retail sales and rising employment should spur retail development. Since 1994, five of the area's regional shopping centers have undergone major renovations, adding food courts and larger department store spaces. Most new construction in the near term will be by big-box retailers developing their own sites.[35]

The Central City

Beginning in the 1970s, progressive planning practices, aggressive initiatives by the Portland Development Commission (PDC), and the philosophy that high-quality, thoughtfully designed public improvements can catalyze successful redevelopment have worked together to bring downtown Portland back from the nearly dead. In addition, Metro—Portland's regional government—has been successful in balancing urban versus suburban interests. This and effective cooperation between the regional government and the business community have been instrumental in the successful development of the central city. The downtown transit mall, which opened in 1978, was a development tool as much as it was a transit improvement.

In 1972, the city approved a downtown plan that clarified the PDC's objectives for downtown development. The overall strategies were to concentrate development on the riverfront and in the downtown core and to activate the downtown by creating a pedestrian environment and mixing commercial, residential, and retail uses. The PDC functioned as the city's "developer," using various financing tools (tax increments, property tax abatements, block grants), design review, and its ability to assemble property as a public entity. Since 1970, more than $1 billion in private funds has been attracted to real estate projects in Portland's new downtown.[36]

The Waterfront and East of the Willamette. Phase I of the plan was to replace Harbor Drive, an expressway running along the waterfront, with a park. In 1974, the highway was removed and in 1977 the 1.2-mile Waterfront Park (renamed Tom McCall Waterfront Park) was completed. Then Governor Tom McCall had transferred interstate freeway funds to build this park (and the downtown transit mall). The area between the waterfront and the downtown core was designated for urban renewal, allowing the city to finance public improvements with tax increment financing. The park's immediate popularity brought a renewed interest in the waterfront and led to developments like RiverPlace.

RiverPlace, a mixed-use development, is part of the PDC's 73-acre South Waterfront Renewal District. Connected to the Tom McCall Waterfront Park, it contains a 74-room hotel, a 200-boat marina, 190 condominiums, 40,000 square feet of office space, retail shops, an athletic club, and 864 parking spaces. The city's active role in its development included assembling the site and issuing tax increment bonds to raise $6 million for street and waterfront improvements. Plans for additional housing, office space, and retail operations are underway.[37]

For most of Portland's history, industrial and transportation uses have occupied (and polluted) the riverfront and kept other uses away. The city realized the potential of the river as a resource for downtown's revitalization, and the recent expansion of recreational, business, and residential land uses onto the banks of the Willamette River clearly illustrates the effectiveness of the city's planning efforts and the key role of transit in Portland.

Across from the downtown on the east side of the Willamette River is the Lloyd Center/Coliseum district, which is planned as the eastern anchor of

the central city. MAX light-rail stations in this area are intended to attract retail and other high-density development like the $262 million, 20,340-seat Rose Garden sports arena, which was completed in late 1995, and the adjacent Oregon Convention Center, which was built by the Metropolitan Exposition-Recreation Commission (MERC).[38] In fact, since MAX opened, the Lloyd Center/Coliseum district has seen $750 million of development activity nearby, and assessed property values in this area increased by 134 percent between 1980 and 1991, compared with a 67.5 percent increase countywide.[39]

Downtown and City Neighborhoods. Downtown and residential neighborhoods were a second target of the PDC-led revitalization efforts and MAX has played a large role here as well. Transportation investments, improvements in pedestrian design, housing initiatives, and historic and natural resource preservation measures were undertaken, all incorporating high-quality urban design elements because, it was thought, they would attract activity and help make the downtown the region's focal point.

In 1978, the city installed a transit mall in the commercial and retail core. It restricted auto access, created exclusive bus lanes, and offered free transit rides within its boundaries. Significant public investments in transit amenities (widened brick sidewalks, trees, transit shelters, street furniture, and art work) were made to encourage private investment. On both the east and west ends of the mall, the PDC built short-term parking garages with ground-floor retail, to compensate for the street parking that had been lost. Downtown zoning was changed to focus the most intense office development along the transit mall.[40]

In 1986, MAX made its appearance on the transit mall. The 15-mile light-rail line ran from downtown through the Lloyd Center district to suburban Gresham. High levels of ridership on MAX, spurred in part by the parking lid (see next section), helped enliven the pedestrian environment and focus development interest on downtown infill and rehabilitation opportunities. Many older office buildings that are adjacent to the transit mall were rehabilitated, and several nearby buildings were converted to housing for seniors.

One of the more notable infill projects along the transit mall was Pioneer Square. Originally a parking block, the site was acquired by the city and turned into a public gathering place in 1984. It has attracted visitors and residents downtown. Two MAX stops make the square easily accessible and much development has taken place around it: shops, offices, the historic Pioneer Courthouse, and other historic building renovations.

Portland lost many of its historic landmark buildings to the wrecking ball after World War II. By the time the downtown plan went into effect, attitudes had changed and historic preservation became a key part of the plan. In 1976, the PDC and the Portland Historic Landmarks Commission established an urban conservation fund. Dozens of historic buildings in the Skidmore/Old Town, Yamhill, and Chinatown districts have been preserved, thereby retaining these neighborhoods' rich mix of architectural styles.

A process of architectural design competitions has helped PDC develop high-quality public environments.[41] Pioneer Square, the Performing Arts Center, the KOIN Center Tower, Pioneer Place, and the Portland Building are products of design competitions—and all are notable landmarks.

These projects also illustrate another important principle of urban redevelopment: total clearance and demolition are unnecessary. PDC's success has come from its focus on small projects and its practice of providing resources, such as loans for historic preservation or housing rehabilitation, for the improvement of the areas surrounding its projects. Perhaps the key to Portland's successful revitalization of its downtown has been PDC's talent for coming up with projects that related to the clear vision and goals of the approved 1972 plan.

Housing has been an important element of the revitalization strategy, and the city has concentrated on the reuse and restoration of existing structures for housing. In order to encourage infill development, increase the supply of affordable housing, stabilize neighborhoods, and bring activity downtown, PDC has established an urban homestead program, a downtown housing preservation program, and a single-family housing loan program.

Parking. Portland has aggressively addressed the parking issue, and used parking regulation as an effective tool for improving air quality, creating a livable and compact urban form, and encouraging transit use.

Parking first became an issue in Portland in the early 1970s, when, under U.S. Environmental Protection Agency regulations for air quality nonattainment areas, actions to reduce vehicle emissions were required. (The metropolitan area was in violation of carbon monoxide standards and ozone standards.) In 1975, the city adopted the Downtown Parking and Circulation Policy (DPCP) establishing a lid on parking and setting strict parking limits for new development.[42] The lid was set at 39,680 spaces. Amendments to the DPCP in 1986 accommodated additional spaces for older and historic buildings. The parking ratio of new office buildings was set at 0.7 spaces for every 1,000 square feet of office space (one-fifth the parking permitted by typical suburban codes). The DPCP also called for the phasing out of surface parking lots in favor of mixed-use developments and parking garages.

The DPCP has had impressive results. A gain of 20,000 downtown jobs from the early 1970s to the 1990s has been accommodated by the addition of only 5,000 parking spaces.[43] The number of surface lot spaces has been reduced from an estimated 15,800 in the 1970s to 11,639 today.[44] The lid has also been instrumental in helping Tri-Met—the regional transit operator—boost the share of downtown workers using transit. Weekday downtown transit trips increased by 30 percent between 1977 and 1988—compared with a 25 percent increase in the number of vehicle trips.[45] The benefit goes both ways: without the transit that Tri-Met provides for downtown employees, downtown would have to add the equivalent of six 42-story parking structures to accommodate commuters.[46] Finally, in respect to air quality, the DPCP also has had good results: the lid in combination with reduced emissions from cleaner-running cars has helped eliminate carbon monoxide violations from downtown since 1984.[47]

However, by 1991, the lid (now at 44,322 spaces) had been reached and the shortage of parking was jeopardizing downtown development. In 1994, the city council unanimously voted to lift the lid and, according to David Logsdon, parking transportation demand manager with the city of Portland, it will probably be officially lifted in 1997 when EPA approves the air quality strategies in the 1995 Central City Transportation Management Plan (CCTMP).[48]

The CCTMP continues the DPCP's ban on surface parking in downtown and extends it to the Lloyd Center district, again in favor of mixed-use development and parking structures. Surface parking is permitted in the central city's industrial areas on the east side of the Willamette River across from downtown—Lower Albina and Central Eastside. The idea behind CCTMP parking rules is to "pinch" the parking supply to encourage multimodel travel.[49]

The CCTMP establishes maximum office parking ratios throughout the central city, ranging from 0.7 spaces per 1,000 square feet of office space in the downtown core to 2 spaces per 1,000 square feet in the Lloyd Center district. The setting of maximum parking ratios for areas such as Lower Albina and North Macadam awaits the completion of the state Department of Environmental Quality's (DEQ) parking ratio program, which was mandated by the legislature in 1993 as part of the state's effort to comply with federal standards for ozone in the Portland region. DEQ parking ratios are not expected to be as stringent as those proposed in the 1995 CCTMP. A third parking program is also in effect—the Land Conservation and Development Commission's Transportation Planning Rule (see "Transportation Planning Rule" in this chapter)—but it generally does not apply to the central city.

Parking is already at a premium throughout most of the central city. Retail and office development has brought drivers downtown, while the supply of parking has remained fairly constant.[50] In 1989, the peak utilization rate for garages in downtown was 80 percent. (The design capacity of a parking garage is generally considered to be 85 percent of its total spaces.) The tightness of the parking market is also indicated by an absence of free parking spaces. Hourly parking is priced to discourage driving and to encourage drivers to park in garages rather than on the street. The hourly rate for street parking downtown was raised to $0.90 in 1995, while the rate at city-owned garages is $0.75. Monthly parking rates in the downtown exceed $100, and few business pay for employee parking.[51]

These parking policies are supported by residents of the region. In a Metro survey conducted for the region's 2040 plan, 17,000 people responded to a questionnaire about land use and transportation issues. Eighty-eight percent were in favor (and 56 percent strongly in favor) of the goal of redeveloping and encouraging growth in the city center; and 54 percent were in favor of reducing the amount of parking—many favoring adding structured parking and others favoring mass transit for accessing businesses affected by a reduction in parking spaces.[52]

In sum, all levels of government have adopted parking policies and measures to address a variety of concerns—air quality, the pedestrian environment, and transit use. The question that concerns developers and others is whether proposed parking policies could hurt development in the central city. Could CCTMP's strict maximum parking ratios thwart central city development and encourage suburban development? Suburban parking is also being regulated. In the words of John Kowalcyzk of the DEQ: "State air quality plans certainly will include a pinch on suburban parking." The question is whether these measures—the DEQ's Parking Ratio Program and the LCDC's Transportation Planning Rule—will be enough to level the development playing field for both the central city and the suburbs. Lenders remain skeptical and view the more liberal suburban parking ratios as a continuing threat to downtown. Tri-Met and Metro are hoping that lowering suburban parking ratios will be enough to enable downtown to remain a competitive location for development.

Regional Government and the Urban Growth Boundary

Metro. In 1973, the state legislature made it mandatory rather than voluntary for the urbanized counties and cities in the Portland region to join the Columbia Regional Association of Governments (CRAG), making Portland one of only three metropolitan areas in the country with a mandated council of governments.[53] CRAG's new structure allowed associate membership by Tri-Met, the Port of Portland, and other adjacent counties and cities.

Another change weighted votes in the body's General Assembly in accordance with member jurisdictions' populations, a change that caused some suburban residents to worry about Portland's dominance. A measure put forth by the Committee to Restore Local Control of Land Planning to abolish all councils of government almost won at the ballot box in 1976.

At the same time, the National Academy for Public Administration awarded the newly created Tri-County Local Government Commission a grant to study the concept of multilevel government in metropolitan areas, and the commission took advantage of this to design a tier of government "that would attend to the common needs of the entire Tri-County community." The commission formally proposed a restructuring of the Metropolitan Service District to combine its regional services with CRAG's planning functions.

The new regional agency would have an elected council, elected executive officers, and an executive planning arm. It would be essentially like Minnesota's Twin Cities Metropolitan Council, with the major difference that its council and officers would be directly elected. In 1978, voters agreed to mandate this regional government by 20,000 votes. Metro was created, the only regional government in the country that is directly elected.

Many mistakes marred Metro's first years. Most of them can be blamed on the inexperience of the regional government's elected leaders and their failures in judging the political thresholds of their constituents. The new regional government tried to do more than was politically possible, but it also scored many successes.

In 1979, Metro created the Joint Policy Advisory Committee on Transportation (JPACT) to meet federal requirements that local general purpose governments participate directly in regional transportation planning. JPACT acted as a forum for elected local officials and transit agency program planners to make regional transportation policy decisions. JPACT and Metro established a working relationship (and combined approval process) that facilitated regional consensus on transportation projects. Although Metro has the power to reject JPACT recommendations, it seldom does, opting instead to work to reach a common agreement.

The building of the Oregon Convention Center in 1986 was another Metro success. Metro created the Metropolitan Exposition-Recreation Commission (MERC) to handle the planning and construction tasks, thereby avoiding the difficulties of convention center planning that many cities have confronted. In addition, when voters approved a $65 million bond issue for the project, Metro's existence—or that of a direct successor—was assured for the next 25 years.

In 1987, executive veto power was initiated at Metro. Metro's executive director replaced the state governor as the person in charge of appointing members of the Boundary Commission. In 1990, voters approved a measure to amend the Oregon constitution to give Metro its own home rule charter. Seats on the Metro Council are strongly contested, viewed by many as being as powerful as seats in the state House of Representatives.[54]

Metro is governed by a 13-member council elected from council districts and a executive officer elected at large. The new government has grown carefully in line with its public support, and the general consensus is that it is a positive force for the region. Because of this support, Metro has been able to take some formidable steps, including the establishment of regional growth boundaries.

The UGB. The creation of Portland's urban growth boundaries (UGBs) began some years earlier. In 1973, the Oregon legislature determined that uncoordinated land use posed a threat to development, the environment, safety, order, and people's convenience, prosperity, and welfare. It also identified a need for properly prepared and coordinated comprehensive plans for cities, counties, regions, and the state as a whole. These findings led to the establishment of the Land Conservation and Development Commission (LCDC) to oversee implementation of a statewide planning program and the creation of the Department of Land Conservation and Development (DLCD) to carry out the directives of the LCDC. The LCDC developed 19 statewide planning goals. Goal 14 called for urban growth boundaries for every incorporated city.

The planning process for UGBs in the Portland region was begun by CRAG in 1977 and completed by Metro. LCDC approved the boundary plan in 1980. This plan coordinated the plans of 24 cities and 60 special service districts, and it was based on projections of the need for urban land through 2000. The boundary has a perimeter of 200 miles and it encloses 234,000 acres (365 square miles). Its two primary and related goals are the preservation of resource lands and the containment of urban development.[55]

All urban areas are contained within the UGB. Rural land has three classifications: exclusive farmland, exclusive forest land, and exception lands. Home-building and other "nonrural" uses are generally not allowed on farmland or forest land. Exception lands are too small or too close to existing development to farm. They can be developed, but only at low, generally uneconomic densities. The UGB was never intended to be static. However, only 2,515 acres have been added to it since 1979, mostly based on a demonstrated need for more urban land.[56]

On the surface, the principal objectives of the state UGB program have been achieved. More than 90 percent of the state's population growth in the 1980s occurred inside UGBs. The state's UGBs have preserved large blocks of rural land for resource use, and resource productivity appears to be increasing. The average size of farms in the Willamette Valley has increased along with productivity per acre. Thus, an area that is home to 80 percent of the state's residents also produces 40 percent of its agricultural products.[57]

Nor, as noted, have UGB restrictions on the supply of land unduly pushed up housing prices. Another beneficial result has been the streamlining of the development and permitting process because of a more predictable planning process.[58]

Not all the results of the UGB have been positive. One unintended consequence has been to make development outside the UGB easier than development within it. Development on exception lands is not subject to the same technical development review as land within the UGB. Residential displacement has been another problem, a result of the infill and redevelopment for higher densities that the UGB encourages.[59]

The biggest problem for Portland is sprawl within the UGB, which is increasing the pressure to expand the boundary. Growth is occurring at only 70 percent of planned density.[60] With Metro forecasting an additional 720,000 residents and 350,000 jobs within the region by 2040, what is the future of the UGB?[61]

Portland's UGB will expand, but where, when, and how are unknown quantities. One of the tasks of the Metro Region 2040 study is to consider expansion needs in relation to demand for growth over the next 50 years. The 2040 plan is under development, the goal being to accommodate Portland's growth for 50 years. It calls for keeping a tight UGB, creating urban communities within the region, focusing growth along transit corridors, preserving open spaces, reducing new lot sizes, and creating compact business areas. Metro has the authority to require changes in local land use and transportation plans to make them consistent with the region's framework plan.

The intention of ongoing planning initiatives in metropolitan Portland is to achieve a clear distinction between urban and rural lands, to put into place an urban to rural transition that lessens sprawl, and to balance redevelopment. The 2040 growth concept seeks to create high-density centers of employment, housing, and transit services. But the region's people have incompatible goals—the preservation of farmland and forest land on the one hand, and low-density neighborhoods and travel by auto—87 percent of trips are by auto[62]—on the other. Ultimately, Metro will have to make development more attractive inside the UGB than outside, if it wants to attain its target densities and accomplish its vision of a compact urban area surrounded by farms and open space.

The Transportation Planning Rule

LCDC's Transportation Planning Rule (TPR) takes a direct approach to the issue of transportation and land use integration. It pledges to ensure that "the state's transportation system supports a pattern of travel and land use in urban areas which will avoid the air pollution, traffic, and livability problems faced by other areas of the country."

This aggressive planning stance won a national award for planning implementation in 1993 from the American Planning Association. When adopted in 1991, the TPR called on local governments to determine if their land use designations, densities, and community design standards support alternative modes of travel and required them to amend development ordinances to encourage infill, cluster development, mixed-use development including housing, and the location of residential development near transit lines.[63]

The TPR, in other words, called for changing current development templates, for coming up with site designs that were less auto dependent. But it did not say how this could be done. There were no generally recognized standards for evaluating whether a development supports alternative modes of transportation. Two initiatives sought to remedy this lack of knowledge. The state Department of Transportation prepared a best management practices handbook and the Oregon chapter of the American Planning Association developed a guide: *Recommendations for Pedestrian, Bicycle, and Transit-Friendly Development Ordinances.*

But business organizations were still concerned about possible impacts of the TPR, and they convinced the LCDC to extend the implementation deadline so that more discussion could take place and some real-world examples be studied. In September 1993, Metro initiated regional discussions on TPR implementation at the request of area planning directors and various business groups. The kickoff meeting of about 60 representatives agreed on a set of objectives for the discussion: examine a number of cases in order to come to a better understanding of the specifics of TPR requirements; explore different ways in which new developments can meet the TPR requirements; and distill some lessons from these examples to serve as guidelines for local jurisdictions.[64]

One example that this group studied was a two-story, 36,000-square-foot suburban office in Gresham, near a MAX transit stop. The original site plan placed the building close to the street, but set back from a future building to take advantage of the proximity to transit and also for better leasing. Parking was at the side and rear. One TPR "makeover" called for relocating entries, paving a path from the plaza to the street, extending pedestrian/bicycle paths into the nearby neighborhood, and adding bicycle parking. A more radical redesign would have rotated the building to face the main street and placed all the parking at the rear. From a development perspective, these revisions caused some problems, including added site and construction costs, a loss of desirable corner spaces, and a less desirable orientation of the offices. This example also taught some lessons about how jurisdictions can encourage site designs that support alternative modes of transportation. Among these was the idea that there should be some flexibility in how buildings can be oriented to the street. The discussants also concluded that projects should be oriented to transit streets, but that trying to orient them to bus transit stops, which can be easily moved, is unwise. And, they felt, the development of standards for pedestrian-friendly environments should be the responsibility of local governments.

The Metro regional discussion group looked at ten projects covering ten different kinds of development. Based on these, it came up with some guidelines, including the following:

- An areawide transportation plan is needed to guide the implementation of the TPR at the site level, a plan that addresses how the site fits into the area's streetscape plan or master plan.
- Standards should be clear, objective, and flexible in order to fit TPR requirements into specific development circumstances. For example, the requirements might specify a maximum building setback from transit streets, with discretion to adjust it within a certain distance.
- The requirements should fit the setting. All areas should support transit, pedestrian, and bicycle facilities, but areas planned for auto-dominated uses should not require the same pedestrian and bicycle facilities as areas that are oriented to transit.
- Regulations should encourage, not discourage, mixed-use projects, which will be difficult to develop.

In April 1995, the LCDC revised the "urban" portions of the TPR, particularly the requirement that new development be more accessible to bicycle and pedestrian traffic and more oriented to transit. Developers welcomed the transformation of detailed standards, as originally proposed, into suggested examples. Another change clarified that the TPR sets minimum, not maximum, requirements. Some local governments would like to adopt more stringent transportation planning standards, and there had been some debate about whether this would be precluded. More specific revisions included the following:

- For buildings at a transit stop, the original requirement that they be located as close as possible to transit was replaced with an option: locate the buildings within 20 feet of a street or provide a pedestrian plaza at the transit stop.
- Communities that charge transportation impact fees must lower fees for projects that are designed for bike, pedestrian, and transit access.
- Cities and counties must minimize their street and right-of-way standards, especially as to width. Besides reducing the costs of local streets, most of which are paid by developers, lower standards make more efficient use of land and reduce travel speeds, which creates an atmosphere that is more conducive to walking and bicycling.

Some advocates of the TPR think that such changes have gutted the original program. Most businesses and developers and some local governments, on the other hand, welcome the changes as a dose of reality that was needed to counteract some heavy-handed and development-insensitive rule making. In any case, the changes demonstrate that there are limits of tolerance for government regulation, even in Oregon, where public sector planning sometime seems to reign supreme.

Recent surveys show that the amount of driving per capita is increasing, and the state's Department of Land Conservation and Development has begun an evaluation of the early results of the TPR.

Outlook: Bringing the Private Sector on Board

Portland's governments—the city, Metro, and Tri-Met—have embraced the principles of livable cities and transit-oriented design. Their thinking has been assisted by a number of leading practitioners of the new urbanism, including Peter Calthorpe and Andres Duany. All but one of the local agencies involved in the planning for the west side rail line have adopted interim development guidelines promoting transit-oriented development. All that remains is for the pri-

Gresham Central

Gresham Central, in downtown Gresham, Oregon, is a 90-unit, market-rate housing complex that was developed by Gresham Development Company, a corporation created by the principals of Westridge Construction at double the density of a typical suburban subdivision on a site carved out of a transit right-of-way. The project, which was completed in 1996, was designed to provide pedestrian links to the Gresham Central light-rail transit station and bus facility.

It involved the first use of federal congestion management and air quality program funding for a joint transit–real estate development project. The Portland regional transportation agency, Tri-Met, obtained permission from the Federal Transit Administration to sell a 0.7-acre parcel to the developer, as part of a development agreement that specified the density, parking, sources of public financing, and the design and required the developer to dedicate a 0.12-acre piece of this parcel for a pedestrian promenade. The developer added this land to its two-acre adjacent site.

Phil Whitmore, Tri-Met's project manager, says that Gresham Central "serves as a model of many of the features that planners in the Portland region think projects should incorporate in order to support transit. Its density—35 units per acre—can support transit. Designed to look more like rowhouses than a monolithic apartment building, it is compatible with its surrounding community. The project provides less parking (1.5 spaces per unit) than comparable suburban developments and it offers good pedestrian connections to transit and other surrounding land uses."

vate sector to get on board, to design and implement the kinds of projects that planners are advocating.

The private response to date has been slow outside the downtown. Many of the projects that have been built along the east line, like the convention center and the arena, have been for the large part public sector deals, or they are located near the transit line just because that is where most of the available sites are located.

Enthusiastic boosters point to more than $1.3 billion of development completed or started adjacent to the MAX line since its opening in 1986. This includes, however, four massive public investments —the $180 million public/private revitalization of Pioneer Place downtown, the $262 million Rose Garden arena, the $85 million Oregon Convention Center, and the $55 million federal building in the Lloyd Center district. For some of these projects, such as the arena, the location on an intown site near transit was an important development consideration.

Recently, more private investment has occurred along the east line, most of it outside downtown. In the Lloyd Center district, the Lloyd Center mall has been renovated and a new hotel has been constructed. Together they add up to an investment of $238 million, and 10 other projects have been developed here.[65] Farther out, the Burnside and Gresham areas have attracted 26 new apartment projects, along with a number of small office and retail projects.[66]

Rudy Kadlub, president of Costa Pacific Homes, points out that the homebuilding industry in Portland consists of many small builders used to building two-story homes on large suburban lots, a product that seems archaic in terms of both demographics (as he notes, only 37 percent of households are traditional families) and planning policy.[67] In 1994, according to Kadlub, higher-density, mixed-use housing communities had a 17 to 19 percent market share nationally, but only a 15 percent share in northwest Portland (and only 5 percent in the suburbs). The home-building market in Portland needs to build for sale housing at higher densities in transit-oriented communities—"different products in different settings." In Kadlub's opinion, making this leap will require innovative builders and innovative local governments.

Tri-Met's efforts to encourage MAX-oriented development began as a behind-the-scenes role. Like the typical transit agency, Tri-Met took the position that its being in the development business was not appropriate. The agency played a Tom Sawyer role, encouraging local agencies to actively promote development near transit and preparing a handbook to show how it can be done. In 1993, Tri-Met decided to become more active, showing by example the types of transit-oriented projects that could be developed. There was ample precedent for this approach in downtown Portland, where the city and Tri-Met had performed extensive streetscaping adjacent to the rail line and transit mall and set high standards for developers to follow. Two significant early projects by Tri-Met in its new activist guise are Gresham Central (see "Gresham Central" feature box) and Beaverton Creek, an ambitious mixed-use project.

Formerly known as Murray West, Beaverton Creek is the first project to be built under Tri-Met's transit-oriented development guidelines, which call for dense, mixed-use development designed for pedestrian access and multiple modes of transportation. The project is located in the upscale suburb of Beaverton, which is in the area known as Silicon Forest where a number of major high-tech corporations have their headquarters. It was planned by a team of landowners—Specht Development, First Western Investments, U.S. Bank, and Tektronix—and Tri-Met. The west side rail line was aligned to go across the property rather than along its boundary (which would have been an easier alignment), in order to maximize the potential for transit-oriented development. The city of Beaverton designated the 120-acre parcel as a

transit overlay zone, which was helpful in the planning while still leaving the tough details to a management committee made up of the four property owners and the transit agency.

A preliminary master plan submitted in March 1995 called for 1,600 multifamily units at densities from 22 to more than 25 units per acre, and 150 to 200 single-family units at a minimum density of 12 units per acre, which is high for suburban Portland.[68] Plans for a mixed-use community center adjacent to the station to the northeast, including an integrated mix of higher-density housing, offices, and transit-oriented retail services, were abandoned when Nike, which owns adjacent property, bought the land for its own use. On the other side of the tracks, approvals for the development of almost 800 medium-density apartments have been obtained and other projects are being contemplated. In April 1996, the city of Beaverton approved Trammell Crow's application for 562 units—garden apartments, townhouses, and one mid-rise building. That same month, the Beaverton Planning Commission held its first hearing on the proposed plan and code amendments for the 120-acre area. Beaverton Creek is a test of the city's new guidelines for development projects. As such, it has revealed some interesting challenges for the transportation/land use relationship, including the following:

Development on a forested site. Had this project not been a high-quality transit-oriented development it may have run into strong opposition. On balance, local residents who were not pleased about losing the forest on the site agreed that this kind of sensitive development was a fair trade.

Density considerations. The proper density for multifamily housing was a matter of some contention between the public sector planners and the developers. The former wanted to push the upper envelope for density, while the developers were concerned that too much density would not be financially feasible or find rapid acceptance in the market.

Public support. As an innovative project in its market, Beaverton Creek has involved more extensive public review, comment, and negotiation than would have been required for a conventional development, as well as transit agency participation. If these kinds of projects are going to catch on they will need not only market acceptance but also strong and dependable public and citizen support, financial as well as political. At Beaverton Creek, public funding will enhance station areas, sidewalks, and landscaping.

Parking standards. Public sector planners wanted to keep parking standards low to encourage transit, and the developers wanted enough parking to meet the needs of their target market. (At Beaverton

Creek, the number of landowners involved made it hard to take advantage of shared parking efficiencies.) At high housing densities—above about 30 units per acre according to Michael Fisher, manager of Tri-Met's land use program—structured parking is necessary, and that is a high-cost item.

None of the developers who survived the recent bad times in Portland wishes to be the first to try out the concept of transit-oriented design. As Kadlub notes, Portlanders' progressivism in politics does not extend into the housing market, where they are pretty traditional consumers. Developers and public sector planners alike think the development industry will become more comfortable with the concept once a few successful projects are built. Tri-Met is hoping that a showcase project will materialize. Until the concept is proven in the market, only projects like Beaverton Creek that have a public sector partner will be undertaken.

To make a dent in the problem, says Henry Markus, station area development coordinator and project manager for Tri-Met, "we need thousands of projects, not one or two." This means making transit-oriented development (TOD) designs, financing packages, and zoning approvals routine in order to keep the soft development costs down. Such an approach seems to fly in the face of housing market critics, who denounce the standard cookie-cutter development model as unimaginative and boring. Arguing that there is a market for TODs even though they are not yet on the market, Markus advocates a checklist type approach to design, financing, and entitlements that would also serve to reassure lenders who are concerned about dealing with untried development concepts.

J. Clayton Hering, of Norris, Beggs, and Simpson, a Portland lender, notes that projects like Beaverton Creek will need credit enhancement and subsidized financing until they have figured out how to adjust to a market in which most people dream of having a two-acre lot. To get to that point, project developers will need flexibility, he says.

"Is there a market for it?" is a critical question for any new land use planning concept. It is really a two-part question that begins: "Can we get developers to build it?" New urbanists, who support TOD and other development strategies that, they hope, will reduce automobile dependency, have been asking that question in Portland for several years. Developers tend to respond: If there is a market for it, we will build it. In some ways, untried TOD might pose a *Field of Dreams* dilemma—if no one builds the field, the old baseball players will not come to play on it.

In an effort to address the issue of what is the market, Metro and others convened a seminar in April 1995 on the economics of mid-rise housing and transit-oriented development. The idea was that comparing residential TODs with a housing product with which developers are familiar would be useful. In most cases, the kind of housing that is proposed for TODs is mid-rise garden apartments that can be either rental or for sale. Developers of TODs in other metropolitan areas participated in Metro's seminar.

Chris Leinberger of Robert Charles Lesser & Company reported that TOD housing typically differs from mid-rise housing in a number of ways: it is typically built at higher densities, development costs are higher and so are rents (and sale prices), and vacancy rates are lower (or sales absorption faster).[69] Higher development costs mean a larger equity requirement in financing, and lower returns than for a conventional project. Because traditional lending institutions tend to be unfamiliar with TODs, developing them involves more shopping for a lender or spending more time with lenders to explain the project.

Can the feasibility gap be closed for TOD projects —the difference between investors' expected financial return and what these projects actually can deliver? The Leinberger case study projects suggest some means of doing this, including the following:

- Public sector grants and subsidies to reduce the net project costs. In the case studies, such financing ranged from 5 percent to 20 percent of total project costs.
- Tax exempt bond financing. Although tax exempt bonds carry a requirement for the inclusion of affordable housing, this low-cost money may offer the best financing alternative for TODs, because often they are located near high-income areas where the inclusionary requirement will have only a small effect on the income stream.
- Flexibility on the part of approval agencies. Regulatory sensitivity to issues that affect site improvement costs, and regulatory flexibility on permitting requirements and development fees can do much to control development costs.
- Construction cost control, to the extent possible. Wood-frame housing costs less than masonry, for example. Janss Court, an unusual project in Santa Monica, California, is a case in point. A three-story wood-frame apartment complex that sits atop a four-story, steel-frame retail/office building was not only substantially less costly to construct but it also achieves a distinctly residential look.[70]

Introducing the topic of parking into a discussion of TOD seems crass. It is like the crazy uncle locked in the basement. No one wants to talk about it, but something needs to be done. Most people living within

Bringing the Mall to the Rail

In 1990, the Winmar Company of Seattle submitted plans for a $100 million, 900,000-square-foot regional mall in Gresham. The mall incorporated a MAX station. It was one of the most innovative proposals for a transit-related suburban shopping mall that has ever been made. The project was dubbed "Operation Break-Even" because the transit agency would, it was believed, receive revenues from the project that would cover its operating subsidy for the entire rail system, making the MAX system self-sustaining in seven to eight years. The rail stop was the focal point of the project, and it would give transit riders an edge over drivers in terms of convenient access. There was one big problem: the project required federal approval because it used federal transportation funding. The debate with the U.S. Department of Transportation over the appropriateness of government funding for this project was protracted, and after uncertainty arose among potential tenants and the recession deepened in 1992, the developer walked away from the project. Who, if anyone, was at fault is still a matter of contention. What the project illustrates is a developer who was willing to develop unconventionally and a public sector that did not deliver on the transportation/development partnership. The project never advanced far enough to discover whether it would gain the support of major retailers or lenders.

---

walking distance of transit still will own cars, even if they do not drive them to work. Transit agencies would like to restrict parking as much as possible, in order to encourage people to rely on transit. Developers would like to have as much parking as will make the residents and business tenants happy.

In downtown Portland, the generally untouchable issue of parking has been addressed head-on. In Portland's suburbs, it is being reconsidered in connection with the parking rule, which requires a 10 percent reduction in per capita parking spaces in 20 years.

In a mixed-use TOD, each different use—residential, office, retail, hotel, cinema, and so forth—has its own parking needs, but some parking can be shared by land uses needing it at different times, such as retail/office space needing daytime parking and residences needing evening parking. It may be possible to lower office parking requirements, for example, through aggressive transportation management programs and parking charges. On the other hand, retailers expect most of their customers to arrive by car and they tend to resist any lowering of parking standards.

A big issue for transit-oriented design is the density at which structured parking is required. When spaces cost between $10,000 and $20,000, parking becomes a crucial cost item, especially for moderate-income housing developments. This tipping point probably varies widely in accordance with site char-

acteristics and the project's design. For general planning purposes, Tri-Met calculates that, at 1.5 spaces per unit, surface parking can accommodate residential densities of up to 24 to 30 units per acre in three-story buildings. Higher densities would require parking structures. Tri-Met thinks that for office buildings, surface parking can accommodate floor/area ratios under 1.0. Higher FARs would require parking structures.[71]

As noted, parking in downtown has been quite constrained, while in the suburbs office developers have been including ample tenant parking at four to five spaces per 1,000 square feet.[72] In terms of transportation purposes, limiting downtown parking is a two-edged sword. One does not wish to make office tenants sufficiently uncomfortable that they move to the suburbs, where transit use is much less heavy. At the same time, higher land values and site constraints in the suburban office market is causing some developers to offer less parking—2.5 to three spaces per 1,000 square feet—and, at some mid- and high-rise projects, even structured parking.[73]

Future reductions in suburban parking ratios may have more to do with public policy than the market. The state Department of Environmental Quality, which in 1993 was directed by the legislature to develop programs to reduce single-occupant commuting trips and new parking spaces for nonresidential developments, says John Kowalcyzk, is likely to recommend a reduction in suburban parking requirements to meet federal air quality standards. And the LCDC's TPR pertaining to reductions in the number of per capita parking spaces will apply chiefly to residential subdivisions.

Lessons

Portland appears to be doing a great deal right to promote high-quality development and efficient transportation, and to avoid the worst kinds of development-induced traffic problems. If it is not doing the best, Portland is certainly doing the most among large urban areas. Unfortunately, it appears that this still is not enough. The growth boundary has not stopped what is derisively referred to as "sprawl;" it has only contained it. The region's emphasis on high-quality transit has so far had little impact on auto traffic and congestion.

What is perhaps most impressive is that, instead of throwing up their hands, public planners and government officials have redoubled their efforts. New policies have been established to focus growth in areas where it is believed it can best be served by public infrastructure, and transportation and other policies have been established to support compact urban growth. There appears to be a solid regional consensus that this is the way to go. What now?

Having deliberately opted for emphasizing transit over highway improvements, Portland is making a "billion-dollar development gamble," the cost of the west side rail project. Because tunneling is required, this line is expected to cost $56 million per mile, well above the price of the original MAX line, which was built mostly on existing rights-of-way. Based on current ridership projections, the west side line will cost $54,000 per daily roundtrip rider.

Such a significant investment makes it all the more important to focus growth, and Portland's light-rail system is being used as a tool to focus investment. The new 2040 regional growth plan is moving forward. The Metro Council adopted the so-called Phase I framework plan in September 1996, and is scheduled to reach a decision on changes to the urban growth boundary by winter 1996. Adoption of the final framework plan is scheduled for the end of 1997, and local compliance plans are expected to be ready by 1998. According to Samuel Seskin, a senior professional associate in the Portland office of transportation consultant Parsons Brinckerhoff, unlike most regional plans, "this one offers no wiggle room. The types and density of uses are tightly drawn. We are entering uncharted territory for American planning."[74]

The public sector has set a strong example, and the big question now is the private sector response. Considerable effort has been devoted to looking at TOD from the perspectives of development markets and developer needs. Several initiatives—like the Beaverton Creek project—in which the public sector is a principal have been undertaken. Advocates of compact development are asking detailed questions about the costs of developing innovative development projects. Many parties, public sector and private sector, are taking care to ensure that public land use and development policies are consistent with what people can afford and are willing to buy.

Portland's cutting-edge experience to date in the area of transportation/development relationships contains some lessons, a few of which follow:

The transportation benefits of improved community designs go well beyond increased transit use. Increased transit ridership is the most readily observable benefit, because data on it is readily available. However, because walking is much more common than transit use, even a slight shift toward more pedestrian and bicycle travel can have a significant impact on driving —and especially on air pollution from driving. Most auto emissions are unrelated to the length of the trip; thus, simply reducing the number of short car trips made can improve air quality more than one might expect. For Beaverton Creek, three times more trips by walking and bicycling than by transit are projected to be made; this number of walking and bicycling trips is three times the number of such trips in more typical suburban communities.

Public policy must be consistent with market needs. This lesson may seems obvious, but it is one that is very difficult to implement. Planning is generally the art of directing growth to where the planners desire it to be and away from where it might naturally gravitate. Succeeding in this endeavor requires a solid understanding of people's locational needs and cost preferences. Portland's public sector planners have become more interested in determining what the market will bear, which, typically, is a private sector concern.

Past successes are no guarantee of future successes. The revitalization of Portland's downtown is an important success story. But the same tools may not work so well in the suburbs. In dealing with complex issues, humility is always a virtue but it is sometimes hard to exercise it when coming from a record of early wins.

Public/private partnerships can prime the pump. If changing the course of certain development patterns is a clear public goal, the public sector may need to become an active partner in development to pull it off. Studies in Portland have shown that certain types of transit-related development that are desirable from a growth policy perspective may not be feasible from a development perspective without some form of subsidy from the public sector.

Notes

1. Douglas R. Porter, "A 50-Year Plan for Metropolitan Portland," *Urban Land*, July 1995, pp. 37–40.

2. Carl Abbott and Margery Post Abbott, *Historical Development of the Metropolitan Service District* (Portland, Oregon: Metro Home Rule Charter Committee, May 1991).

3. Larry W. Price, ed., *Portland's Changing Landscape* (proceedings of a meeting of the Association of American Geographers, Portland, Oregon, April 22–26, 1987), p. 42.

4. Portland Development Commission, *Moving into the Fourth Decade* (Portland, Oregon: Portland Development Commission, 1989), p. 7.

5. Price, p. 42.

6. As quoted in Donald Canty, "Portland: How Its Downtown Became the Lazarus of American Cities," *Architecture*, July 1986, p. 32.

7. Ibid.

8. Abbot and Abbot, p. 11.

9. Ibid., p. 16.

10. Sonny Condor, *Metro Measured*, Metro Occasional Paper Series, no. 3. (Portland, Oregon: Metro Planning Department Data Resource Center, 1994).

11. Robert C. Brannan, Ethan Seltzer, and Michael A. Wert, "Coordinating Portland's Urban Growth Plan and the Western Bypass Study," in *ITE 1991 Compendium of Technical Papers* (Washington, D.C.: Institute of Traffic Engineers, 1991), p. 68.

12. From data compiled by Metro Planning Department Data Resource Center, June 13, 1995, based on Metro and U.S. Bureau of Economic Aanalysis employment data.

13. *2040 Framework Update* (Portland, Oregon: Metro, spring/summer 1995), p. 3.

14. Portland Development Commission, *Portland Facts* (Portland, Oregon: Portland Development Commission, June 1994), pp. 3, 5.

15. Data from Portland Metropolitan Association of Building Owners and Managers, *1994 Portland Metropolitan Office Guide* (Portland, Oregon: Portland Metropolitan Association of Building Owners and Managers, 1994).

16. Grubb & Ellis Research Services, "Portland Metropolitan Area," in *ULI Market Profiles: 1994* (Washington, D.C.: ULI–the Urban Land Institute, 1994), p. 125.

17. Ibid., p. 128.

18. Data from *Viewpoint 1994: Real Estate Value Trends* (Valuation International, Ltd. Valuation Network, 1994) as cited in Grubb & Ellis (see note 16).

19. *1994 Portland Office Guide* (see note 15), p. 28.

20. Grubb & Ellis, p. 128.

21. E.D. Hovee & Company, *Portland Central City Transportation Management Plan: Development Case Studies*, preliminary draft (prepared for Portland Department of Transportation, December 1991), p. 7.

22. *1994 Portland Office Guide*, p. 60.

23. E.D. Hovee, p. 13.

24. Shiels & Obletz, *Portland Central City Transportation Management Plan: Development Case Study* (prepared for Portland Department of Transportation, 1991).

25. E.D. Hovee, p. 6.

26. Grubb & Ellis, p. 128.

27. Developers report that in downtown the cost of constructing a parking space, excluding land, runs more than $10,000 in above-ground structures (the range is $6,000 to $20,000); $15,000 in below-ground structures; and $2,500 to $3,500 on surface lots. Outside downtown, structured parking reportedly costs $8,000 or less per space. See E.D. Hovee, p.15.

28. Grubb & Ellis, p. 127.

29. *Portland Facts* (see note 14), p. 22.

30. J. Thomas Black, "Rising Residential Land Prices," *Urban Land*, July 1996, p. 26.

31. *Clayton-Fillmore Report*, March 1994, p. 5.

32. *Portland Facts*, p. 7.

33. Ibid., p. 7.

34. See "Pioneer Place," *Project Reference File*, v. 25, no. 4, January–March 1995 (ULI–the Urban Land Institute).

35. Grubb & Ellis, p. 127.

36. *Moving into the Fourth Decade* (see note 4), p. 4.

37. See "RiverPlace," *Project Reference File*, v. 18, no. 3, (ULI–the Urban Land Institute, January–March 1988).

38. Portland Office of Transportation, *Central City Transportation Management Plan* (Portland, Oregon: Portland Office of Transportation, April 1995), p. 81.

39. G.B. Arrington, Jr., "Beyond the Field of Dreams: Light Rail and Growth Management in Portland," in *Proceedings of the Seventh National Conference on Light-Rail Transit* (Washington, D.C.: National Academy of Science Press, 1995), pp. 9–10.

40. Rick Gustafson, "Private Sector Role in Developing Portland's Light Rail Station Sites," in *Public/Private Partnerships in Transit*, v. 2 (Washington, D.C.: U.S. Department of Transportation, April 1985), p. 88.

41. Price (see note 3), p. 43.

42. CCTMP (see note 38), p . 5.

43. Ibid., p. 5.

44. Gordon Oliver, "Portland Revs Up for Action," *Planning*, August 1994, p.12

45. JHK & Associates, *Portland Central City Transportation Management Plan: Technical Analysis* (Portland, Oregon: Portland Department of Transportation, January 1992), table 4-7.

46. Arrington (see note 39), p. 43.

47. William Loudon, E. Coleman, S. Iwata, and R. Gustafson, "Changing Transportation Policy in the City of Portland," in *1994 Compendium of Technical Papers* (Washington, D.C.: Institute of Transportation Engineers, 1994), pp. 512–516.

48. David Logsdon, interview by author, June 30, 1994.

49. CCTMP, p. 39.

50. JHK & Associates, p. 8.

51. *Metro Region 2040 Update* (Portland, Oregon: Metro, fall 1994), pp. 4, 15.

52. Information from a workshop that ULI convened in Portland, September 20, 1994, to examine the private sector interest in transit-oriented design.

53. The information on CRAG and Metro in this section is largely from Abbot and Abbot (see note 2), pp. 13–29.

54. Ibid., p. 31.

55. Arthur C. Nelson, "Blazing New Planning Trails in Oregon," *Urban Land*, August 1990, p. 32.

56. V. Gail Easley, *Staying Inside the Lines: Urban Growth Boundaries*, Planning Advisory Service Report no. 440 (Chicago: American Planning Association, 1992), p. 18.

57. Nelson, p. 34.

58. Easley, p. 18.

59. Nelson, p. 34.

60. Tom Walsh, "There Is a Bus or Train in Your Future," *1994 Portland Metropolitan Office Guide* (see note 15), p. 22.

61. *Metro Region 2040 Update* (see note 51), p. 5.

62. Portland Metro, unpublished data from a 1994 household survey.

63. "Oregon's Transportation Planning Rule," *Planning*, March 1993, pp. 13–15.

64. See Cogan Owens Cogan, *Transportation Planning Rule Implementation Regional Guidelines* (report prepared for Metro, February 1994).

65. Arrington (see note 39).

66. Ibid.

67. ULI workshop, September 20, 1994 (see note 52).

68. Fletcher Farr Ayotte, *Beaverton Creek: A Residential Mixed-Use/Major Employment Center Project* (report prepared for Beaverton Creek Master Plan Committee, March 1995).

69. Christopher B. Leinberger, "Developer Roundtable: Case Studies Exhibit Packet," (paper presented at a seminar, "The Economics of Mid-Rise Housing and TODs," sponsored by the Oregon Mortgage Bankers Association and the West Side Light Rail Station Community Planning Committee, April 28, 1995, in Portland, Oregon).

70. "Janss Court," *Project Reference File*, v. 21, no. 3, (ULI–the Urban Land Institute, January–March 1991).

71. Michael Fisher (land use program coordinator for Tri-Met), interview by author, May 23, 1995.

72. E.D. Hovee (see note 21), p. 9.

73. Ibid., p. 9.

74. Samuel Seskin, interview by author, October 1, 1996.

Chapter 5

Transportation and Development in Atlanta
Coming to Grips with Prosperity

You can make yourself grow if you want, but urban problems come with urban prosperity.

— James Carson, Atlanta developer

Location and transportation have always been the drivers of Atlanta's growth. From its first days as a settlement clustered around a rail intersection, Atlanta has been profoundly shaped by its central location, its transportation network, and the penchant of its citizens for economic progress. The historic city was shaped by the railroad network and Atlanta's inner suburbs were shaped by a streetcar system. Development of Hartsfield International Airport as a major hub made Atlanta the transportation center of the Southeast. And development of an extensive freeway system made it the economic center of the Southeast.

Atlanta has made itself burgeon. But, as Atlanta developer James Carson, who is active in transportation and growth issues, has said: "You can make yourself grow if you want, but urban problems come with urban prosperity." As an expanding and prosperous metropolitan area, Atlanta is just beginning to come to grips with some of these problems.

The 1996 Summer Olympics, expansions of the highway and transit systems, and a robust economy are some of the regional high points of recent years. The picture was not always this bright. The city of Atlanta was hit especially hard by recession in the 1970s and experienced the fifth largest population loss among U.S. cities. However, the history of Atlanta illustrates its ability to recover from setbacks. Between 1980 and 1994, according to the Atlanta Regional Commission (ARC), the ten-county ARC region's jobs increased by 74 percent. ARC forecasts that the region's population will increase from 2.6 million in 1993 to 3.1 million by 2000.[1]

This growth has not come without a price. Atlanta's suburbs are thriving, but its central city has suffered and its role in the region has declined. Other serious problems in the region include traffic congestion, a heavy dependence on cars, and a growing disparity between opportunities for whites and opportunities for blacks.

Before the 1970s: Laying the Foundation

Atlanta came to be because the Georgia legislature sought to establish a railhead that would connect the state's Atlantic ports with the western United States.[2] In 1937, a stake marking the Western and Atlantic Railroad's zero-mile post was driven into the ground. The first railroad station was built in 1842, and development soon clustered around it. The settlement's name was Terminus. Soon additional rail links formed an emerging rail network, providing jobs to support the growing population. In 1847, the thriving town was incorporated as Atlanta.[3]

During the Civil War, Atlanta was more a boom town than a city and it suffered severe destruction when General William T. Sherman's troops set it ablaze in 1864. Despite this devastation, Atlanta forged ahead and quickly became a symbol of the New South. Residents actively promoted the city and its commerce and industry, partly through hosting several expositions on the themes of regional commerce and regional economic self-sufficiency. Capitalizing on its transportation advantage over neighboring cities, Atlanta soon emerged as the chief distribution center in the Southeast.[4]

By 1885, Atlanta was a flourishing city of 75,000 residents. Streets were laid out in a grid along the railroad tracks that surrounded and bisected the city. The first commercial centers grew up around rail stations. However, the emerging urban environments, with their noise, dirt, and bustle, were sometimes more suitable for commerce than for living, and people began eyeing the surrounding woodlands as more residentially attractive.

Suburban development began in 1889. Joel Hurt developed Inman Park east of the city as the first planned suburb for downtown workers. The Atlanta and Edgewood Street Railroad Company provided streetcar service between the new suburbs and the downtown business district. A 1908 plan by Frederick Law Olmstead for a suburban development at Druid Hills east of the city featured Atlanta's first parkway. With streetcars and a parkway, the move to the suburbs was well underway.[5]

Meanwhile, two Atlantas—white Atlanta and black Atlanta—began to emerge. The post–Civil War black presence in Atlanta has been strong, with blacks constituting 40 percent of the city's population in the years from 1870 to 1900. After the Civil War, black businesses along Decatur, Marietta, and Peachtree streets were a prominent feature of the CBD. But a color line was drawn after a major race riot in 1906 and the rise of the Ku Klux Klan to a position of political power.[6] (In 1923, the Klan had 15,000 members in Atlanta, and it established its headquarters here.) Atlanta's black population and black-owned businesses were pushed out of the CBD. They moved to the east side of the city, with the businesses reestablishing themselves in the Auburn Avenue business district and the residents moving to the Old Fourth Ward, a neighborhood that is still predominantly black and was once home to Martin Luther King, Jr.

Often called "the Atlanta spirit," the city's elevated sense of civic pride has helped it survive several difficult periods—including the aftermath of a second fire (1917) that destroyed 73 blocks. At about the same time, after numerous lynchings, Atlanta acquired a reputation as a city where racism and violence were more than tolerated.[7]

In 1925, business leaders and the chamber of commerce responded with the Forward Atlanta campaign. The program's goals were to encourage businesses from outside the region to relocate in Atlanta, foster the expansion of local firms, and make improvements in education, transportation, and urban design. A lead-off $1 million advertising campaign was followed by a blitz of civic investments, among which was the construction of New Union Station (dedicated in 1930) to replace the 1905 main railroad station, and by a plan to beautify the city with plantings, scenic walkways, and grand plazas.

An ambitious plan to improve downtown's street network started in the 1920s and continued through the 1950s. Atlanta underground was created by con-struction of an elevated street network above the rail tracks and the relocation of retail operations to the new (upper-level) street level. (In the 1960s, the rediscovered lower level of buildings was turned into an entertainment district that was short-lived, a victim of population decline, recession, the liberalization of suburban liquor laws, fears about crime, subway construction, and management problems. In the 1980s, a second attempt to redevelop the underground worked: using $145 million in revenue bonds backed by the city, the Rouse Company turned the space into a festival marketplace, which has become a major attraction.)

In 1929, Atlanta councilman William B. Hartsfield lobbied the city to buy a dirt racetrack south of the city for a municipal airport.[8] This decision established Atlanta's air industry, which has played an important and continuing role in the regional economy.

A second generation of suburbs began opening up in the 1920s. New roads and more cars made it possible for people to move to locations further outside the city where streetcars did not reach, and the affluent flight to the suburbs was on. White residents moved north to Ansley Park and Buckhead. Black residents moved to the western suburbs.

Boom turned to bust in the early 1930s. The Forward Atlanta campaign came to an early end and the economy stagnated. Many local leaders thought the adjustment to a peacetime economy after World War II would be difficult. No development had taken place since the depression and many of the city's neighborhoods were in a bad state of neglect. The revitalization strategy that Atlanta adopted balanced two goals: rebuilding the central city and providing transportation to make the suburbs mobile.

Regional Mobility

In 1946, voters approved a bond issue for almost $20 million to implement a road plan that had been designed by H.W. Lochner and Company. The Lochner plan predated the initiation of the U.S. interstate highway system by a decade. The foresight of Atlanta's business leaders, politicians, and planners in this instance paid off in terms of positioning the region perfectly for federal funds for highway construction when they became available.[9]

In that the CBD was the region's only large center of employment, the Lochner plan focused on moving commuters from the outer suburbs into the central city. A system of spokes (I-75, I-85, and I-20) radiated from the central city. (Except for the addition of the perimeter highway, the current freeway system still closely resembles the original Lochner plan.)

A new era was ushered in and the city's current political boundaries were established in 1952 with

Figure 5-1
Population of Atlanta, 1970–1995

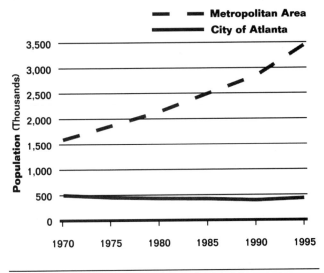

Sources: Decennial Census; and Atlanta Regional Commission.

the Plan for Improvement, a program that was supported by the city's mayor, William Hartsfield, and that sought to spur investment in the central city and the suburbs simultaneously.[10] Atlanta annexed the surrounding counties, thus increasing in size from 37 square miles to 118 square miles and gaining 100,000 new residents.

In 1947, officials from DeKalb County, Fulton County, and the city of Atlanta joined together to create the Metropolitan Planning Commission, the first publicly supported multicounty agency in the country.[11] The MPC later became the Atlanta Region Metropolitan Planning Commission, which became, in 1971, the Atlanta Regional Commission.

Central City Renewal

Atlanta's urban renewal efforts began in the 1950s. As in many American cities, they involved the demolition of many historic structures in the name of progress and the wholesale clearing of entire blocks —neighborhoods and business districts—for new uses: stadiums, civic centers, parking lots.

The first skyscraper in 25 years—the Fulton National Bank Building—went up in 1955. In the following years, more office buildings were developed, some of them outside Five Points, the historic business district. Development at Peachtree Center, five blocks to the north, and along the radial perimeter highway signaled the beginning of Atlanta's shift to a polycentric office market.[12]

Thinking that Atlanta's changing development patterns should be better directed, Mayor Ivan Allen, Jr.,

helped launch a second Forward Atlanta campaign in 1960. The campaign's intensive promotion of Atlanta attracted 173 new plants, 323 new out-of-town firms, and more than 70,000 new jobs.[13]

The second Forward Atlanta also pushed a number of key construction projects: the completion of the expressway system, the construction of a new auditorium and stadium, and the creation (by the state's General Assembly) of the Metropolitan Atlanta Rapid Transit Authority (MARTA). In the 1970s, the central city underwent a decade of economic decline. Fortunately, the road and transit projects started under the auspices of Forward Atlanta had laid a foundation for the city's recovery in the 1980s.[14]

Current Conditions and Recent Initiatives: Decentralization and Investment in Transportation

Atlanta Compared with Other Metropolitan Regions

Atlanta's unrestrained growth is reflected in its driving.[15] Atlantans drive an average of 34.9 miles a day, giving them a commanding lead in the mileage department over drivers in other urban areas.[16] Georgia's low gas tax—the second lowest in the country—also contributes to high VMT.[17]

According to the Texas Transportation Institute, Atlanta ranks tenth out of 50 cities in terms of highest traffic congestion in 1992—near New York, Houston, San Diego, and Seattle. More importantly, congestion in Atlanta is growing at a faster rate than in most cities. Among the 50 cities, its percentage increase in congestion between 1982 and 1992 was seventh highest, following Sacramento, Minneapolis, San Francisco, Columbus, Salt Lake City, and San Diego. The annual cost of congestion in Atlanta is estimated to be $1.1 billion.[18]

Despite having 1.01 freeway miles per 1,000 residents—the second most extensive freeway system among 40 metropolitan statistical areas (MSAs)—Atlantans spend a lot of time in their cars. As a percentage of all work trips, single-occupant vehicle work trips increased from 58 percent in 1970 to 78 percent in 1990. Currently, the Atlanta metropolitan area has a 90 percent level of auto dependence for work trips. The length of work trips has increased as more suburb-to-suburb commutes are made, many of them on arterial highways and interstates. In 1990, about 53 percent of travel on main roads was on freeways. In 1990, commuting time in Atlanta averaged 26 minutes, which was the fifth highest time among 32 MSAs. Commute times were comparable

Figure 5-2
Transportation Indicators for Atlanta, 1970–1990

	1970	1980	1990	Percent Change 1970–1980	Percent Change 1980–1990
Population	1,597,816	2,138,231	2,833,511	34%	33%
Commuters	660,013	950,030	1,481,781	44	56
In Private Vehicles	562,253	836,837	1,344,050	49	61
Driving Alone	475,377	732,353	1,242,028	54	70
Passengers	86,876	116,808	102,022	34	–13
By Transit	52,218	70,165	67,958	34	–3
Commute Share					
Transit	8%	7%	5%	–7	–38
Private Vehicles	85%	88%	91%	3	3
Travel					
Daily Vehicle-Miles Traveled (Thousands)	n/a	23,500	71,501	n/a	204
Annual Transit Trips (Thousands)	48,346	120,160	147,882	149	23
Median Single-Family Lot Value	n/a	$13,250	$18,000	n/a	36

Sources: Decennial Census; *Journey-to-Work Trends in the United States and Its Major Metropolitan Areas, 1960–1990* (Federal Highway Administration, no. PL-94-012); and ULI residential land price surveys.

in Houston (26 minutes), Los Angeles (26 minutes), and San Francisco (25 minutes).[19]

Having only 1,550 persons per square mile, Atlanta is one of the least densely populated urban areas. Population density is comparable in Houston (1,810), Phoenix (1,980), and Cincinnati (2,010).[20] Between 1970 and 1990, the density of the central city decreased, while the densities of the urban core counties, the suburban counties, and the fringe counties all increased.[21]

A striking trait of the Atlanta region is the small size of its central city—in terms of both area and population—relative to the whole region. In 1990, when the ARC region had only nine counties, the city of Atlanta had a 16 percent share of the regional population and occupied 5 percent of the region's total land area. The average central-city share of regional population among the largest 47 metropolitan regions was, by comparison, 33 percent. The city of Atlanta's small size poses a dilemma: if 84 percent of the region's residents live in the suburbs and if 72 percent of its workers have jobs in the suburbs, how can any policy adopted by the city of Atlanta have an impact on the people and economic activity of the region?[22]

Growth in Atlanta is well above metropolitan averages. In a ranking by *American Demographics* magazine of the top U.S. metropolitan areas in terms of population growth, income growth, and density

between 1970 and 1990, eight of Atlanta's counties made the top 50 list. Fayette County was first in the nation. The other counties making this list were Gwinnett, Cherokee, Forsyth, Paulding, Henry, Rockdale, and Douglas.[23] Median household income in the Atlanta region was $38,816 in 1989, well above the national median of $30,056.[24] Nationally, the proportion of middle-class families is declining; in Atlanta, it is growing.[25]

Development Market Trends

The Atlanta MSA is composed of more than 20 counties covering 5,121 square miles. The smaller ten-county ARC planning district—the ARC Atlanta Region—contained 2,558,000 people in 1990, which is 82 percent of the MSA population. For the past quarter century, the MSA and the ARC region have experienced explosive growth in population and jobs. At the same time, the city's population decreased from almost 500,000 in 1970 to 394,000 in 1990.

Annual population growth in the ARC region over the last decade has averaged 66,000.[26] Most of the population growth that has occurred is from people moving to the Atlanta MSA from outside it.[27] The ARC predicts that the city's population will increase moderately through 1995 but still not reach 1980 levels, while the suburban population will continue to grow rapidly. About 54,300 jobs will be added annually through 2000, according to an ARC forecast.

By 2000, the region is expected to have 3.1 million residents and more than 1.9 million jobs.[28]

Atlanta's economy is highly diverse. Growth is not dependent on one or two key industries and, while recessions like the one in 1990 can slow down growth, the economy tends to recover quickly. In 1994, employment was divided among the various major industry sectors as follows: services (27.5 percent); retail trade (18.2 percent); government (13.8 percent); manufacturing (10 percent); wholesale trade (9.2 percent); transportation, communications, and utilities (8.9 percent); finance, insurance, and real estate (7.6 percent); and construction (4.1 percent).[29]

The region's four largest employers are as follows: Delta Airlines (26,000 employees), AT&T (20,000), Bell-South (16,000), and Lockheed (11,000). Other companies with a sizable presence include IBM, Nations-Bank, the Coca-Cola Company, UPS, and Turner Broadcasting.[30] Small and medium-sized companies are an important element in the economy. Compared with other large metropolitan areas, Atlanta has a low rate of business failures, a high number of incorporations, above-average personal income, a low cost of living, and a large number of support programs for small businesses.[31]

Expansion of the economic base has continued into recent times. Almost 100,000 jobs were created in 1994, and it was estimated that 94,000 jobs would be added in 1995. On the eve of the Summer Olympics, the regional unemployment rate was a low 4.5 percent, almost a full point below the national rate.[32] Job growth will probably fall off after the Olympics, which have provided an enormous stimulus to the economy.[33]

The Olympics brought an estimated $10 billion into the economy, created 80,000 temporary jobs, and left many civic improvements, including a 60-acre park on a downtown site that was mostly abandoned buildings and parking lots and an 85,000-seat stadium that will be the home of the Atlanta Braves. Some critics, however, feel that the city lost an opportunity afforded by the Olympics to make long-term improvements and address some of the larger urban challenges facing it.[34]

Like many metropolitan areas, Atlanta is growing outward in what critics have called the "unlimited low-density vision."[35] Also in line with many metropolitan areas, Atlanta contains a "favored quarter" —in its case lying north of the city—where most of the region's jobs and upper- and middle-end housing are located.[36] According to many observers, some of the social and environmental consequences that are associated with a low-density pattern of growth include the dominance of the automobile, scattered low-density workplaces, the isolation of inner-city residents, the emergence of socially and economically homogenous neighborhoods, and the breakup of the government structure into small, nearly autonomous local governments.[37] Metropolitan Atlanta exhibits all of these characteristics to one degree or another.

The history of development in Atlanta provides an excellent example of what Christopher Leinberger has described as a succession of generations of metro cores in different, increasingly distant locations.[38] In Leinberger's model of metropolitan development, the first generation metro core is the old city. Second generation cores—suburban in character but located within central city beltways—developed in the 1960s but tended to falter later as their office space became obsolete and nearby residential neighborhoods declined. Third generation cores outside the beltways developed in the 1970s and 1980s. They contained a good mix of upper- and middle-income housing to support the office labor force and, by the 1980s, many of them had more occupied office space than the region's downtown. The 1980s boom spawned a fourth generation of farther out metro cores. Development of these cores tended to overbuild their markets, but a number of them have since been able to recover.

In general, in Atlanta, the city and inner suburbs— the first generation and second generation cores— lack the necessary stability of employment and middle-income residents to attract further investment. In 1983, downtown Atlanta contained almost one-third of the region's occupied office space; just ten years later, this share was down to one-quarter. The Northeast Expressway, a second generation core, was capturing 12 percent of the region's occupied office space in 1983; by 1993, this area contained only 7 percent of the region's total. Without employment gains and the improvement of their residential bases, Atlanta's downtown and second generation cores are in danger of further deterioration.

Atlanta's prosperous third generation metro cores stand in sharp contrast to downtown and inner-suburb decline. Some, like Buckhead/Lenox and Perimeter North (which although not exactly outside the perimeter highway is a third generation core), have matured and become more urban in form. Many are connected to mass transit and have a good mix of retail uses and higher than average housing densities. The safe "urban experience" that they offer is attractive to many people, and their market support has helped them to establish strong economic bases that make further development likely.

Atlanta's fourth generation cores—Georgia 400, Marietta Town Center, and Oakwood/Gwinnett— advanced from a zero share of regional office space in 1983 to 12 percent in 1993. These cores have a

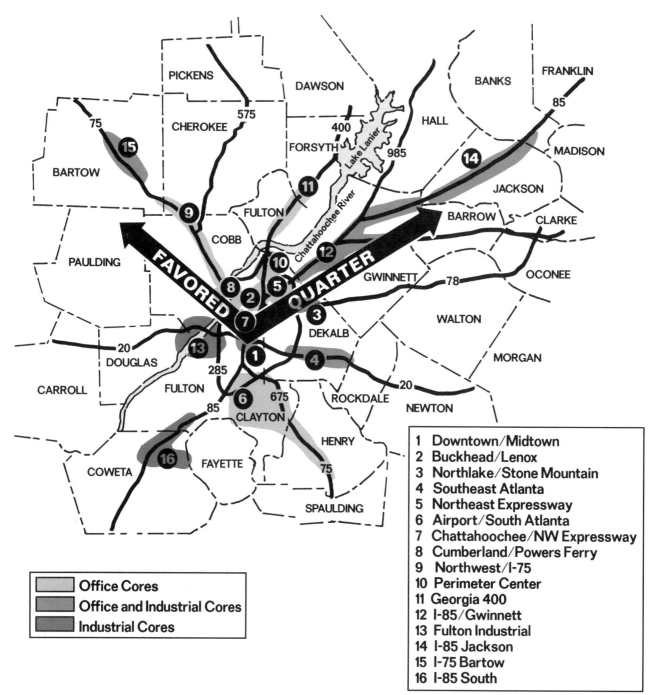

Location of Atlanta's metro core areas.

Robert Charles Lesser & Co.

Legend:
- Office Cores
- Office and Industrial Cores
- Industrial Cores

Core areas:
1 Downtown/Midtown
2 Buckhead/Lenox
3 Northlake/Stone Mountain
4 Southeast Atlanta
5 Northeast Expressway
6 Airport/South Atlanta
7 Chattahoochee/NW Expressway
8 Cumberland/Powers Ferry
9 Northwest/I-75
10 Perimeter Center
11 Georgia 400
12 I-85/Gwinnett
13 Fulton Industrial
14 I-85 Jackson
15 I-75 Bartow
16 I-85 South

number of development advantages: an abundant supply of land, few social (crime or welfare) problems, generally progrowth attitudes, and lower than average tax burdens. Since 1983, almost 80 percent of Atlanta's net employment growth has taken place in third generation and fourth generation metro cores located in the northern suburbs.[39]

Although few fourth generation metro cores have run out of land, Leinberger sees the recent emergence of fifth generation cores located generally seven to 20 miles beyond the fourth generation cores. The development of these cores is being driven by the relocation of companies rather than by population growth. Housing follows jobs to these locations. Dallas and Houston provide the best examples. In Atlanta, development in this pattern is taking place along I-75 and I-85, 45 to 70 miles north of downtown. If present trends continue, says Leinberger, Atlanta's expansion "will eventually reach the South Carolina border."[40]

Population Location Trends. The decentralization of population in Atlanta is based on a number of factors. Federally funded freeways and other infrastructure development have opened up the suburbs, and federal tax policies have encouraged the dispersal of residential and office development. Atlanta lacks natural barriers such as mountains or rivers to its spreading out. A strong market preference for large-lot, single-family housing has pushed development outward.

While the city of Atlanta's population declined 7 percent between 1980 and 1990, the fastest-growing counties were registering large increases: Fayette (114 percent), Gwinnett (111 percent), Cherokee (74 percent), Henry (61 percent), and Cobb (50 percent).[41] Figure 5-3 shows the region's high-growth areas (in terms of population and employment) by Census tract between 1990 and 1995. With the exception of two tracts in Henry County, the areas with the fastest-growing population are located outside the perimeter highway, in the northeast and northwest corridors along Georgia 400 and I-85. Most tracts within the perimeter highway lost population. According to ARC forecasts to 2000, some tracts just outside the perimeter highway will start losing population; the northern suburbs (Gwinnett County along I-85, Cherokee County along I-575, and Cobb County along I-75) will continue to attract the most growth; and some southern counties (Henry, Rockdale, and Clayton) will attain population growth levels rivaling those of their northern counterparts.[42]

Employment Location Trends. Employment trends essentially have mirrored population trends. One follows the other, but the penchant for jobs to follow people is greater than it is for people to follow jobs.[43] Many companies have moved to the suburbs to take advantage of lower land costs, lower labor costs, and lower property taxes and insurance premiums.

The strongest employment growth has been in the northern suburbs, which increased their share of the region's jobs from 29 percent in 1970 to 52 percent in 1990. With an increase of 24,500 jobs, Fulton County was the region's fastest-growing county, followed by Cobb (20,900 added jobs) and Gwinnett (16,150 jobs).[44] Most Census tracts showing employment losses are located within the perimeter highway.

The city of Atlanta's share of regional employment in 1990 stood at only 29 percent, down from 55 percent in 1970. Retail businesses left the city first, in the 1960s and 1970s. Office-based businesses and corporate headquarters began leaving in the 1980s.[45] Despite these losses, the CBD has been able to survive by becoming the prime location for hotels and conventions, and by attracting jobs in the government and producer services (legal, accounting, and advertising) sectors. Three major universities downtown are a strong economic presence.

Suburban centers—most of them located near the exits and major intersections of the I-285 beltway—have captured most of Atlanta's employment growth. Typically auto-oriented, these centers have an employment mix similar to downtown's and they sometimes rival downtown in size. Among the largest of these centers, shown in Figure 5-4, are the following:

- The Perimeter Center/Georgia 400, with 19 million square feet of office space (which is 3 million more than in the CBD);
- Cumberland/Galleria, with 18 million square feet of office space;
- Buckhead/Lenox, with 9 million square feet; and
- Midtown, also with 9 million square feet of office space.[46]

The Office Market. The possible construction of an additional loop around the perimeter and the strong job growth expected through the end of this century should keep demand for suburban office space high, if the suburbs can accommodate it.

The Atlanta office market is finally close to recovery after the combination of overbuilding in the 1980s and the recession in the early 1990s. The amount of excess Class A space is declining. Among several new build-to-suit facilities that were constructed in 1995 are a 650,000-square-foot Hewlett-Packard facility in the Perimeter Center area and a 377,000-square-foot Home Depot headquarters in Cobb County. Only 400,000 square feet of speculative office space was on the market at year-end 1995.

In 1994, the downtown office vacancy rate climbed to 22.6 percent.[47] For the first time, rental rates in some suburban office centers surpassed rents found in downtown.[48] Vacancies have since abated. In mid-1995, the vacancy rate for Class A downtown space was 13 percent and for suburban space it was 7.6 percent.

The Housing Market. Atlanta has been one of the nation's top-performing housing markets for a number of years. In the 20-county region, almost 200,000 housing permits were issued between 1990 and 1995. During this time period, 65 percent of housing construction took place in five counties: Cobb, Clayton, DeKalb, Fulton, and Gwinnett. Almost 85 percent of the permits were for single-family detached houses.[49] This strong preference for single-family houses stems from the strength of the economy, low interest rates, and a continuing influx of corporate transferees.[50] The median price of residential lots grew 39 percent between 1990 and 1995; but compared with other metropolitan areas, lot prices in

Figure 5-3

High-Growth Areas in Atlanta, 1990–1995

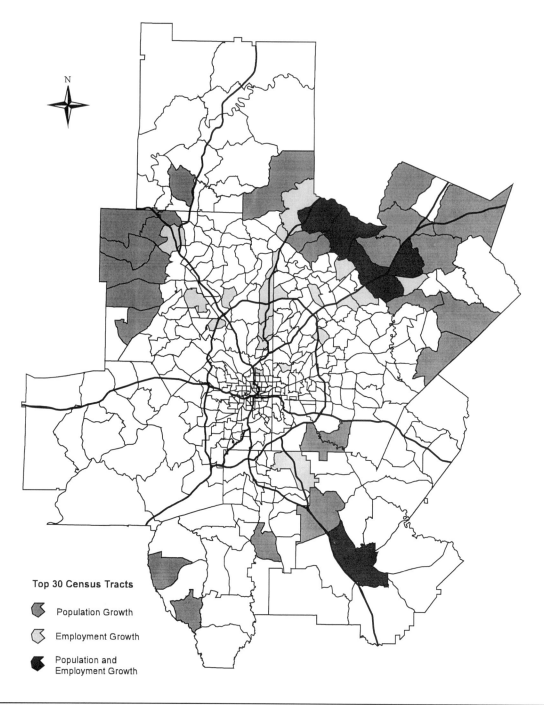

Top 30 Census Tracts

◆ Population Growth

◆ Employment Growth

◆ Population and Employment Growth

Source: Atlanta Regional Commission.

Atlanta are low, ranking 26th out of 30 metropolitan areas examined.[51]

Freeways

Outward expansion has necessitated a considerable investment in roads and created a heavy reliance on private autos. The freeway system, except for the perimeter highway, essentially has followed the 1946 Lochner plan (see earlier "Regional Mobility" section).

The downtown connector freeway (I-75 and I-85) started construction in 1951 and opened for traffic in 1964. By 1969, more than 122 miles of interstate freeway were in operation. Within six years after it opened, the downtown connector was serving 93,000 vehicles per day (vpd), 80 percent more than had been projected initially. By 1980, its vpd had increased to 135,000. With only two lanes in each direction, the freeways could not handle this volume of traffic.

In 1992, the Georgia Department of Transportation (DOT) initiated a $1.4 billion Freeing the Freeways expansion program. It called for improvements on 122 miles of mostly older, four-lane freeways. With the program's expansion of four-lane freeways to eight lanes and ten lanes, freeway lane-miles increased from 500 (in 1980) to 1,400 in 1990. Peak hour speeds increased by 10 miles per hour, shortening travel times, reducing accident rates, and improving air quality.

By 1991, Atlanta's freeways were carrying from 50 percent to 100 percent more traffic, without congestion. The massive freeway expansion had nearly tripled road capacity and doubled traffic volumes, without, as some had feared, worsening air quality. In fact, in large part because of the improved efficiency of traffic flows (helped, of course, by the stricter emissions standards that had come into effect for new cars), the region lost its status as a nonattainment area for carbon monoxide pollution. The more efficient traffic flows also produced a sharp 20 percent reduction in the accident rate, which translated into an annual cost savings of $19 million. In the view of the Georgia DOT, the Freeing the Freeways program demonstrated conclusively that expanding freeway capacity was a viable option for handling transportation needs in Atlanta.[52]

However, it was more a reaction to contemporary congestion than it was a plan for the future, and its positive initial results were short-lived. Continued growth in traffic in the central area and rapid increases in suburban traffic have eroded the savings in travel time, and regional traffic congestion, as was shown in Chapter 1, was worse in the early 1990s than in the 1980s.

In 1993, the Georgia 400 was completed. (Its official name is T. Harvey Mathis Parkway, after the developer who was the driving force behind its construction.) Going directly from the northern suburbs to downtown, this highway has opened the Buckhead area for development. A toll is charged on the Georgia 400 segment from I-285 into the city.

The Georgia DOT has set aside $85 million to add six lanes to a 3.2-mile segment of the eight-lane I-85 in Gwinnett County, plus room for future HOV lanes. A number of additional arterials and capacity expansion projects are being undertaken to accommodate suburb-to-suburb commutes, including the Ronald Reagan Parkway, Route 120, Route 138 through

Figure 5-4
Office Inventory in Atlanta by Major Employment Center, 1990

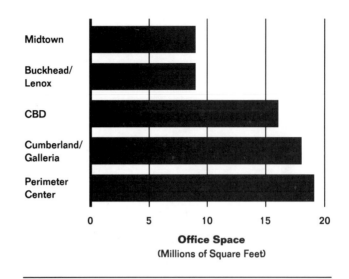

Source: *The Dynamics of Change* (Research Atlanta, Inc., 1993).

Clayton, Henry, and Rockdale counties, the east/west connector in Cobb County, and I-20.[53]

Despite these improvements, the road system's capacity is inadequate. The Georgia DOT reported a 20 to 30 percent increase in traffic on radial freeways between 1990 and 1994.[54] Metropolitan area traffic may double by 2010, according to some estimates.[55]

Atlanta's traffic congestion is not a problem that lends itself to a quick-fix solution. Current policies will have to be changed and politically difficult decisions will have to be made.

MARTA: The South's First Subway
By the early 1960s, planners had realized that an automobile-dominated transportation system would not be able to adequately carry the region's future rush hour radial loads.[56] Based on an MPC plan that included a rapid-rail transit system, the state General Assembly created the Metropolitan Atlanta Rapid Transit Authority (MARTA) in 1965. The agency started out operating a fleet of buses. Its first plan for a heavy-rail system was rejected by voters in 1968, who objected to the use of property taxes to finance it, the cost uncertainties, and the lack of public involvement in planning decisions.[57]

A new plan involving 53 miles of rail lines and eight miles of exclusive busways was submitted, allowing for greater public participation and proposing a sales tax financing scheme.[58] Voters approved it in 1971. By 1979, the system's east line extended from

Figure 5-5

Annual Transit Trips in Atlanta, 1980–1994

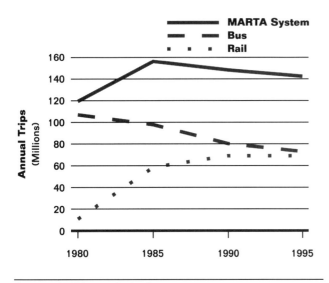

Source: MARTA annual ridership reports.

Georgia State University to the Avondale station near Decatur and its west line ran from Five Points to Hightower. In 1988, the rail reached Hartsfield International Airport, making Atlanta one of the few regions with a direct rail connection to its main airport. There were 38.2 miles of track in 1995, which increased to 46 miles in mid-1996.[59]

A 1995 study of patronage found that rail ridership since the early 1980s has increased on the north/south line more than on the east/west line. The north/south line covers more territory and serves the region's rapidly expanding northern tier, from which comes most of the region's new transit ridership.[60] And this line connects to the airport, which has become an extremely popular MARTA destination, in the south.

Figure 5-5 provides a history of transit ridership in Atlanta since the introduction of rail. The addition of rail gave transit ridership an initial boost. Total annual trips increased substantially between 1980 and 1985, but bus trips dropped precipitously. After 1985, rail ridership leveled off and, in 1994, it reached a plateau of around 70 million annual trips. Bus ridership continued to decline until it reached a similar level in 1994. Despite increases in population in MARTA's service area, total transit ridership has declined steadily since 1985.[61]

Transit is also losing mode share. Transit trips capture only a small and declining share of the commuter market. The solo driving share has increased

steadily over time (from 85 percent, in 1980, to 90 percent, in 1990), while transit's share has been declining (from 7.4 percent, in 1980, to 4.7 percent, in 1990).

Many factors have contributed to the decline in transit's share of work trips. One is the suburbanization of the work force. Between 1980 and 1990, the percentage of the region's workers who lived in the central counties declined from 25 percent to 21 percent. This shift of the work force to the suburbs may seem slight, but it compounds the push toward solo driving that is the result of the city's lessening importance as an employment center and the emergence of multiple suburban employment centers with limited or no transit services.

Fixed-rail systems have a major disadvantage in growing regions: rail facilities cannot easily be matched to employment and population growth.[62] MARTA's radial rail system focuses on the downtown and does not adequately serve the region's growing suburban centers. One transit agency response has been to expand park-and-ride facilities at newer stations on the north line, enabling nearby suburbanites to drive to the stations, park, and ride to destinations served by rail. MARTA has also rerouted its bus system from a radial arrangement with downtown as the hub to a concentric arrangement that provides better cross-suburban connections.

MARTA rail, it should be noted, is a young system that has not yet reached its prime commuter markets. When they were at the same stage of development, neither Washington, D.C.'s Metro nor San Francisco's BART had reached their major edge cities.[63] Even before the mid-1996 extension of the north line, MARTA's volume of weekday ridership reached 90 percent of daily riders carried by San Francisco's BART.

In June 1996, an extension of the north line brought rail into Buckhead and Perimeter Center, with stations at Buckhead, Medical Center, and Dunwoody. Running along the median of the Georgia 400, the extension brings MARTA to the Atlanta Financial Center, Capital Center Plaza, and several hospitals.

Has MARTA rail influenced development patterns? Transit-oriented development in rail corridors has been difficult to achieve. In some cases, local governments have used zoning to preserve the residential character of the neighborhoods around stations. Three suburban counties (Gwinnett, Cobb, and Clayton) rejected MARTA service altogether. Much of the south line runs next to busy rail lines and near manufacturing and warehouse centers, which makes most development along it impossible or not feasible.[64]

A number of large office buildings were built in midtown as a consequence of the the city's creation in 1973 of special public interest districts that

Figure 5-6

Transit versus Solo Driving in Atlanta, 1960–1990

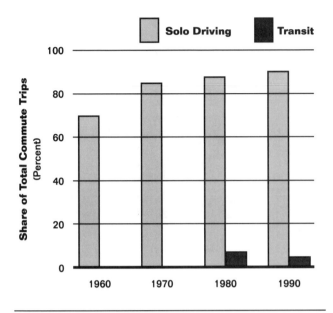

Source: *Journey-to-Work Trends in the United States and Its Major Metropolitan Areas, 1960–1990* (Federal Highway Administration, no. PL-94-012).

allowed higher development densities around MARTA stations.[65]

MARTA likes to think that the rail system has catalyzed development. One transit agency official once noted: "the existence of a MARTA line contributes to development; [but] it does not control it."[66] A 1988 study by Cambridge Systematics—which included an evaluation of MARTA's development impacts based on data from extensive interviews—found that "case studies furnish no evidence that rail transit has shaped regional land use. . . . [There is] mixed and/or modest evidence of rail transit's ability to shape development near downtown and other commercial station locations." The evidence for transit-related development in Atlanta is incidental, states the study's report, which goes on to say that "the sense of disappointment and naiveté is clearly evident regarding the planners' expectations of the formative power of transit." Richard Courtney, chief land use planner for the Atlanta Regional Commission, reports that as of late 1996, "MARTA has had no overall effect on redirecting regional growth patterns. There are some specific stations, such as the Arts Center north of downtown, where intense development would not have happened without MARTA, but such examples are few. Sprawl continues in Atlanta."[67]

Outlook: A New Need for Regional Planning

Atlanta's transportation network has made the region's extraordinary growth possible. The extensive highway system has allowed goods to flow efficiently throughout the region and people to commute long distances to work.[68] Hartsfield International Airport, the second busiest airport in the world, has a new $305 million international terminal, and it continues to be a major generator of economic activity in the region.[69]

Abundant locational options and low transportation costs create the competitive land markets and low land prices that are a big part of what attracts businesses to Atlanta. Regions with extensive transportation and communications networks and few limits on urban expansion enjoy a competitive business advantage. Among the consequences are a lower cost of living and higher wages.[70]

But Atlanta did not plan its low-density, dispersed growth. The region's form comes from economic forces and a general progrowth attitude, not deliberate planning practices. The region has successfully embraced growth and economic development as a vision. In the process, it has become, without question, the dominant economic center of the Southeast. It does, however, still face challenges. Chief among these are environmental quality, balanced growth, and changing travel patterns. Regional coordination would seem to be an important element in Atlanta's ability to respond effectively to these challenges.

Air and Water

Air quality and water quality are problems. Atlanta is a nonattainment region for EPA air quality standards, and its ozone level was listed as serious in 1990. Development has caused a deterioration in the region's water quality. Fifty streams do not meet water quality standards, and the contributing factor for 90 percent of these was found to be urban runoff.[71] Millions of dollars in fines have been paid to federal and state environmental agencies because of jurisdictions' failure to take adequate action on the management of stormwater runoff.

Water supply is another looming concern, especially for the southern part of the region. At one point, ARC was predicting that the regional water supply would be exhausted at the 3.5 million population mark. Now the agency says that, because of improvements in technology and changes in how water is used, the supply will last for at least 25 years—when population is expected to be 4.17 million.[72] John Huey, the Atlanta-born managing editor of *Fortune* magazine, thinks Atlanta may find a way out:

"The only thing that will stop [Atlanta from growing] is if it runs out of water, and then it'll probably go steal it from someone else, like L.A. did."

Balanced North/South Growth

To the north are the affluent neighborhoods, the white neighborhoods, most of the new office and retail development, and, since 1980, three-fourths of the region's population growth and four-fifths of its employment growth. To the south (and west) are African-American communities and Hartsfield International Airport, which has not been strong as a development catalyst. Public investment policies have tended to exacerbate the north/south disparity. Even in the 1960s, when the perimeter highway was built the northern part was constructed first, giving the north a jump-start on the rest of the region.[73]

Many of the problems affecting the northern suburbs—traffic congestion, inadequacy of infrastructure, land use conflicts—could be relieved if growth were more evenly distributed around the region.

Growth imbalances have also made access to jobs difficult for African Americans. In 1990, African Americans made up 43 percent of the total population of the southern suburbs, up from 33 percent in 1980. In the same time period, the share of the region's total jobs that are located in the southern suburbs declined from 20 percent to 19 percent.[74] A criticism that has been leveled at MARTA is that it concentrates on luring affluent, white, north suburban residents to transit while it neglects to provide adequate levels of service to potential customers in southern and western suburbs.[75] As job opportunities continue to disappear from traditional African-American communities, MARTA's service priorities and highway planning will become more important for the residents of these communities.

Keeping Up with the Commute

Atlanta's transportation network is designed to handle the traditional suburb-to-city commute, which is being made by an increasingly smaller share of workers, while an increasingly larger share is making either city-to-suburb or suburb-to-suburb commutes. MARTA's transit system is focused on Fulton and DeKalb counties, while Cobb County has a separate transit service. Transit service is not even provided in the remaining ARC counties, which account for an increasing share of growth.

Widening the northern stretch of the perimeter highway might help it handle the cross-county commute and alleviate congestion, but it would also further encourage solo driving. Every time expanded road capacity saves driving time, MARTA's compet-

Figure 5-7

Vehicle-Miles Traveled in Atlanta And Other High-Growth Regions, 1990

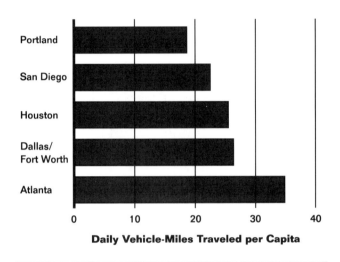

Daily Vehicle-Miles Traveled per Capita

Source: 1990 Highway Statistics (Federal Highway Administration).

itive edge is reduced. Adding an outer loop or at least a partial loop in the northern counties is an idea that is under active consideration.

One outer loop project that has been studied extends 211 miles around Atlanta, 20 miles beyond the I-285 perimeter highway, and is estimated to cost between $2.16 and $5 billion.[76] It would cut through stretches of rural areas, including many environmentally sensitive areas and historically significant sites. An early study predicts it would encourage the development of scattered activity centers at major interchanges. It would also put considerable infrastructure and service demands on the local governments of the jurisdictions through which it cuts.

A partial loop, the northern arc, is a more feasible option that is also under consideration. The arc would run from U.S. 411 at I-75 in Bartow County to the Georgia 400 in Forsyth County, and possibly extend to Georgia 316 in Gwinnett County.

The full outer loop proposal has generated much opposition from environmental groups and homeowners. The major concerns are uncontrolled growth and environmental harm. One local journalist has said that he fears it will solidify Atlanta's "manifest destiny" growth policies. MARTA's board has opposed it, calling it a spur to the continuation of transit-unfriendly land use patterns.[77] If the outer loop plans turn into pavement, fifth generation metro cores would appear in some of the region's most rural areas. The Georgia DOT defends the project as necessary to accommodate the amount of population and economic growth that is forecast for the region. If it is not built, says the DOT, the economy of the entire state will be affected.[78]

The northern arc option lacks a well-defined intent. Is its purpose to relieve congestion on the perimeter highway? Or is it to facilitate development farther north of the city? Many Atlantans feel the main benefactor would be developers and landowners, not the public, and the arc's current design tends to support that view. Only a small segment of the arc goes through an urbanized area, but more than 30 exits are planned—most of them going into areas that are largely undeveloped. A positive note is that the rights-of-way are planned to accommodate transit, either an exclusive bus lane or a rail line.[79]

The ARC has money in its 2010 Regional Transportation Plan (RTP) and its 1995 Transportation Improvement Program (TIP) for preliminary studies on sections of the outer loop. No money has been dedicated for construction.

The 1995 TIP focuses on keeping ongoing projects moving forward and on improving air quality. The plan contains 30 new projects designed to address air quality issues and no projects that add to the highway system's capacity to accommodate single-occupant vehicles. ARC proposes spending $1.25 billion in matched federal funds. The TIP sets aside $675 million for expanding the transportation system as follows:

- $294 million for the construction and expansion of roadways;
- $219 million for transit and alternative modes of transportation;
- $125 million for a multimodal terminal downtown; and
- $37 million for bicycle and pedestrian facilities.

The TIP also sets aside $360 million for managing the existing highway and transit systems and $203 million for maintenance of the highway system.

Funds for constructing HOV lanes are included in the TIP. Atlanta's past efforts to relieve congestion through HOV lanes have fallen short, with the Georgia DOT having used federal money to construct such lanes in the 1980s but having quickly turned them into extra highway lanes. Twenty-four-hour HOV lanes were constructed on most of the interstate highways inside the I-285 perimeter highway in order to move the traffic expected for the Olympic Games; they will, it is hoped, continue to be used as HOV lanes in order to encourage carpooling.

A proposal from the State Passenger Rail Authority to provide commuter-rail service from some outlying communities including Athens, Gainesville, Cobb County, and Cherokee County, would not address the cross-county commute, but would provide some benefits to the communities served. The proposal has run up against CSX reluctance to permit passenger service on its heavily used rail lines, but CSX might look upon it more favorably in the future as the company looks for additional sources of revenue from its operations.

As MARTA nears completion of its heavy-rail system, it will become less of a construction authority and more of an operating authority. Since the cost of building rail extensions is so high, the authority's efforts will shift to the development of areas around existing stations. MARTA has developed a new program, Transit Oriented Design Livable Communities, and it has begun working with local governments to carry it forward. The authority is currently working on the creation of a development partnership for land it owns around the Lindbergh station, where MARTA's headquarters is located on the north line from downtown.

Regional Coordination

All these challenges—air and water quality, balanced growth, and effective transportation planning—lend themselves to regional solutions. Thus, a necessary starting point is improved coordination between the region's local governments. ARC is the approved metropolitan planning organization for Atlanta, but although ISTEA gives it some authority on issues involving federal funding, it has little power to implement its plans. A fundamental problem is that ARC has jurisdiction over only ten counties in a region that is composed of 20 counties.

Without a regional process for guiding land use and transportation decisions, Atlanta will find solutions to its growth-related problems difficult to find. Local governments are still trading off regional impacts for small projects in fear of losing their power base, tax base, and ability to attract businesses.[80] The handling of difficult interjurisdictional issues like air quality and transportation planning will require the state of Georgia to take on greater responsibility for regional planning issues.

Lessons

Among regions whose vision is essentially unlimited and low-density growth, Atlanta is a prime and dynamic example, and it also is experiencing most of the unwanted consequences of such growth—environmental problems, jobs/population imbalances, and traffic congestion. Despite such problems and the criticisms of some observers, the region continues to manufacture jobs, which serve as a strong

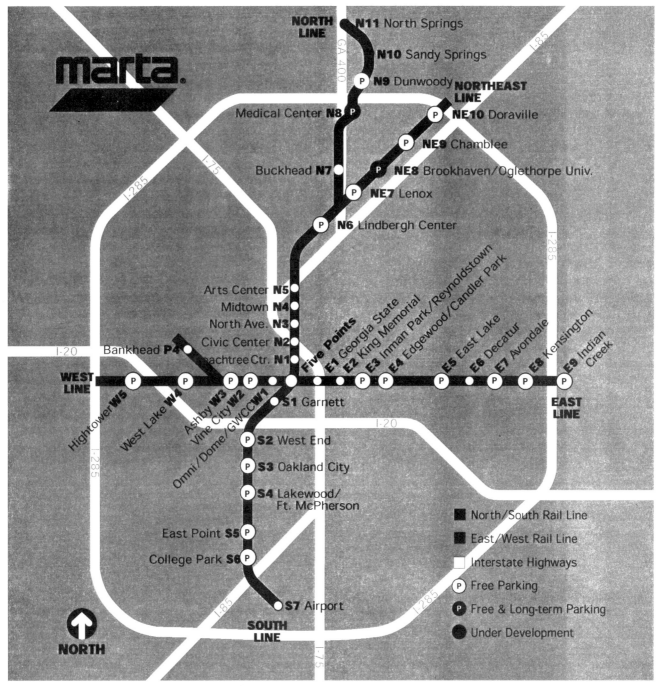

MARTA rail system.

magnet for in-migration from throughout the South and farther away.

But Atlanta's traditional boosterism may be challenged in its effort to continue to support unconstrained growth with transportation facilities, as suggested by the difficulties that supporters of an outer perimeter highway have encountered. The region may be approaching some limits in its ability to continue to grow in an unlimited fashion, both economically and geographically. Moving forward, Atlanta will need to learn three important lessons: collaborative regional planning, revitalization of existing communities and infrastructure, and long-term infrastructure planning.

It will take true regional planning to accommodate growth in an efficient manner. The Atlanta region has a long history of regional planning, which has led to the building of an excellent freeway system, the development of one of the most effective new rapid-rail transit systems in the United States, the protection of the Chattahoochee River corridor, and the provision of other regional amenities. This long planning tradition, however, has not led to a true cooperative approach to managing growth and the impacts of growth. Cities and counties have competed for the development that accompanies each successive wave of outward urbanization, viewing growth as a sign of arriving rather than as a regional management challenge. The highly regarded Atlanta Regional Commission has no real control over its members and it now represents only ten counties in a region with twice that number. In addition, the role played by the city of Atlanta continues to diminish, as suburban growth draws people and jobs away from the central city and inner suburbs.

Existing road and transit capacity can be better used. Each outward wave of development has left behind the "debris" of neglected and even abandoned real estate and infrastructure, and an accumulating set of social and financial problems. While outward growth creates a need for major investments in new infrastructure, it leaves behind excess capacity. A focus on building up the older parts of Atlanta—which is already happening in a few areas—to accommodate some future growth would make better use of existing roads and transit services. It would also create communities in which residents' transportation requirements could be satisfied without the extraordinary levels of driving that are normal in Atlanta.

Accommodating growth through community revitalization would also help Atlanta to develop a transit market from within. Atlanta's transit system is one of the most highly regarded in North America, and its new rail system is one of the most successful—in total ridership as well as in terms of serving people who need transit the most. However, MARTA rail and bus ridership peaked in the mid-1980s, and has been slipping. (For the 17 days of the 1996 Olympics, ridership soared to over 17 million, a number that is normally served in a period three times as long.) Adding riders by building new rail lines is increasingly expensive. But bringing people to live and work in areas that already have good rail service adds new transit riders for very little additional cost.

Keeping the regional transportation system in condition to support a dynamic economy and population requires consistency of infrastructure investment and an effective constituency for growth. Transportation improvements in Atlanta have tended to be a matter of boom or bust. The region's ad hoc, solve-problems-as-they-arise mentality needs to be replaced with an ongoing, reliable infrastructure investment program. The Freeing the Freeways program, the $1.4 billion effort launched by the Georgia DOT in 1982 to expand I-85 and I-285, is an extraordinary example of an effort to revitalize aging roads, but it needs to be more than a one-time effort. Infrastructure stands in need of constant improvement.

A sustained improvements program appears to require a constituency for growth. All the key infrastructure providers—highway and transit agencies, cities, counties, and the state—need to recognize that maintaining adequate transportation involves a long-term commitment, and that plans for transportation improvements must be developed and carried out regularly and consistently. The successful staging of the 1992 Olympic Games for a world audience is the latest, but by no means the only, example of Atlanta's ability to pull together when circumstances demand it. Management of the region's future growth will require nothing less.

Notes

1. Atlanta Regional Commission, *Atlanta Region Outlook* (Atlanta: Atlanta Regional Commission, February 1995), p. 20.

2. Tom Walker, "Atlanta Strives to Stay Brave and Beautiful," *Planning*, April 1989, p. 7.

3. Dana F. White and Timothy J. Crimmins, "How Atlanta Grew: Cool Heads, Hot Air, and Hard Work," in *Urban Atlanta: Redefining the Role of the City*, ed. Andrew Marshall Hamer (Atlanta: Georgia State University, 1980), p. 28.

4. Andrew Marshall Hamer, "Urban Perspectives for the 1980s," in *Urban Atlanta* (see note 3), p. 4.

5. White and Crimmins, p. 30.

6. Ibid., p. 31.

7. Ibid., p. 32.

8. Walker (see note 2), p. 8.

9. Ibid., p. 8.

10. White and Crimmins, p. 36.

11. Pat Murdoch, "Grow Now, Worry Later," *Planning*, April 1989, p. 22.

12. White and Crimmins, p. 37.

13. Ibid., p. 39.

14. Walker, p. 9.

15. Patrick Vicors (Atlanta Regional Commission), interview by author, October 20, 1995.

16. Mary K. Teets, *1993 Highway Statistics* (Washington, D.C.: U.S. Department of Transportation, 1993), p. V-72.

17. Jeff Rader (Atlanta Chamber of Commerce), interview by author, April 2, 1996.

18. All the data in this paragraph are from *Trends in Urban Roadway Congestion—1982 to 1992* (College Station, Texas: Texas Transportation Institute, 1995), pp. 12, 27.

19. Michael A. Rossetti and Barbara S. Eversole, *Journey-to-Work Trends in the United States and Its Major Metropolitan Areas, 1960–1990* (Washington, D.C.: U.S. Department of Transportation, 1993), p. 4-38.

20. U.S. Federal Highway Administration, *Highway Statistics: 1990* (Washington, D.C.: U.S. Department of Transportation, 1991), table HM-72.

21. Truman Hartshorn and Keith Ihlanfeldt, *The Dynamics of Change: An Analysis of Growth in Metropolitan Atlanta over the Past Two Decades* (Atlanta: Research Atlanta, Inc., 1993), p. 13.

22. Anthony Downs, remarks as cited in *Key Aspects of the Future of the Atlanta Region* (report of the ULI Atlanta District Council meeting, February 1994), pp. 2–3.

23. G. Scott Thomas, "America's Hottest Counties," *American Demographics*, September 1991, p. 36.

24. *Atlanta Region Outlook* (see note 1), p. 17.

25. Hartshorn and Ihlanfeldt, p. vi.

26. *Atlanta Region Outlook*, p. 29.

27. Hartshorn and Ihlanfeldt, p. 8.

28. *Atlanta Region Outlook*, p. 3.

29. Ibid., p. 43.

30. Ernst & Young, "Atlanta," *National Real Estate Index: Quarterly Market Reports*, 3rd quarter 1994, p. 3.

31. Ibid., p. 3.

32. Rader (see note 17).

33. Ernst & Young, p. 2.

34. Ken Friedlein, "Lost Opportunity in Atlanta?" *Architectural Record*, August 1994, p. 71.

35. Anthony Downs, "Some Controversial Aspects of the Atlanta Region's Future," *The Brookings Review*, summer 1994, p. 27.

36 . Henry L. Diamond and Patrick F. Noonan, *Land Use in America* (Washington, D.C.: Island Press, 1996), p. 210.

37. Ibid., p. 210.

38. Christopher B. Leinberger, "The Changing Location of Investment and Development Opportunities," *Urban Land*, May 1995, pp. 31–36.

39. Ibid., p. 36.

40. Christopher B. Leinberger, "The Changing Location of Investment and Development Opportunities," *Black's Guide*, 1995, p. 41.

41. From data compiled by the Atlanta Regional Commission, August 10, 1995, from the 1990 Census.

42. *Atlanta Region Outlook*, p. 36.

43. Hartshorn and Ihlanfeldt, p. 59.

44. *Atlanta Region Outlook*, p. 41.

45. Hartshorn and Ihlanfeldt, p. 59.

46. Ibid., p. 64.

47. "1994 Atlanta Review," *Southeast Real Estate News*, March 1995, p. A-7.

48. David Greenfield, "City Review: Atlanta," *National Real Estate Investor*, May 1994, p. 80.

49. Hammer, Siler, George Associates, "Atlanta Metropolitan Area," in *ULI Market Profiles: 1996* (Washington, D.C.: ULI–the Urban Land Institute, 1996).

50. Downs, *Key Aspects* (see note 22).

51. J. Thomas Black, "Rising Residential Land Prices," *Urban Land*, July 1996, p. 26.

52. Georgia Department of Transportation and Georgia Division of the Federal Highway Administration, *Freeing Atlanta: More Capacity Proves Viable Strategy for Congestion Relief* (Atlanta, Georgia: Georgia Department of Transportation, 1991), p. 2.

53. Rader (see note 17).

54. Jeff Rader (Atlanta Chamber of Commerce), interview by author, August 17, 1995.

55. David Goldberg, "DOT Defends Outer Loop," *Atlanta Journal and Constitution*, November 7, 1995, p. 1.

56. Walker (see note 2), p. 8.

57. Georgia Institute of Technology, *MARTA Impact Study: Level of Service, Operating Economics, and Social Benefits* (Atlanta: Atlanta Regional Commission, June 1995), p. 3.

58. Ibid., p. 5.

59. U.S. Department of Transportation, *Transit Profiles of the Thirty Largest Transit Agencies* (Washington, D.C.: U.S. Department of Transportation, December 1994).

60. Georgia Institute of Technology, *MARTA Impact Study: Patronage Analysis* (Atlanta: Atlanta Regional Commission, June 1994), p. 8.

61. *MARTA Impact Study* (see note 57), p. 24.

62. Ibid., p. 52.

63. Ibid., p. 1.

64. Bert Roughton, Jr., "MARTA: So Far, So Good," *Planning*, April 1989, p. 16.

65. Ibid., p. 15.

66. Ibid., p. 17.

67. Richard Courtney (Atlanta Regional Commission), interview by author, September 12, 1996.

68. *Atlanta Region Outlook*, p. 11.

69. In 1993–1994, Atlanta surpassed Dallas/Fort Worth in the number of passenger emplanements. See Marilyn Gross and Richard N. Feldman, *National Transportation Statistics: 1996* (Washington, D.C.: U.S. Department of Transportation, Bureau of Transportation Statistics, 1995), p. 83.

70. See J. Thomas Black, "The Economics of Sprawl," *Urban Land*, March 1996, p. 6.

71. *Atlanta Region Outlook*, p. 21.

72. Jim Auchmutey, "Growing Pains: Is Atlanta Getting Too Big?" *Atlanta Journal and Constitution*, December 31, 1995, p. D-1.

73. James Carson (chairman of CARTER), interview by author, April 2, 1996.

74. Hartshorn and Ihlanfeldt, p. 79.

75. Roughton, p. 14.

76. Hartshorn and Ihlanfeldt, p. 79.

77. Editorial, *Atlanta Journal and Constitution*, December 29, 1995, p. A-12.

78. Jerry Pachucki (director of planning and development for MARTA), interview by author, April 2, 1996.

79. Goldberg (see note 55), p. 1.

80. Pachucki, April 2, 1996.

Chapter 6

Transportation and Development in Phoenix
Urban Villages in the Desert

If the people who sputter about sprawl paused long enough to listen, they'd realize that we are all concerned about the same issues. . . . They'd admit that the plans were in place for subdivisions like mine long before we knew we were moving here.

— Elizabeth Farquhar, *Arizona Republic*

Phoenix is the epitome of a post–World War II, western, auto-oriented community. Located in the aptly named Valley of the Sun, the Phoenix region has experienced substantial growth in the postwar years. Many newcomers are seeking a definite lifestyle. More than one in four families has a swimming pool and one in five adults owns a recreational vehicle.

The area's recipe for growth seems simple, although some of the ingredients may be hard to come by: "Start with the Phoenix metropolitan area's warm climate, plentiful jobs, low cost of living, powerful developers, and land speculators. Add palm trees for good measure, and you've got a recipe for the premier growth region of the West."[1]

The city of Phoenix is the capital of Arizona and the population center of Maricopa County, a growing metropolitan region of 2.5 million people. Phoenix is the seventh largest city and the region is the 17th largest urbanized area in the United States. Several municipalities in the metropolitan area have over 150,000 people.

It would be hard to overstate the region's growth. Phoenix is the only large metropolitan area to rank among the three fastest-growing U.S. regions in each of the three decades beginning with the 1960s. In 1950, the central city had slightly more than 100,000 people; in 1990, it had almost 1 million—an 819 percent increase in just four decades. In 1950, the city fit into 17 square miles. Today it spreads over 400 square miles. Its average density fell 63 percent between 1950 and 1990—the greatest drop in density of all Sunbelt cities except Las Vegas.[2]

Phoenix is a study in contrasts. In a city dominated by the auto, the freeway share of trips is very small and voters have twice refused to finance the construc-

tion of additional freeways. In the ultimate suburban community, planners are promoting an urban village concept that celebrates mixed land uses and higher densities than are usually found in southwestern communities.

Phoenix presents other anomalies. Deep suspicion (if not worse) greets all proposals to regulate housing and growth, yet many residents have willingly bought homes in private communities that impose extraordinary restrictions on everything from the vegetation they may plant to how long their guests can stay. Phoenix sits in the middle of the 400-square-mile Sonoran desert, but much of its lifestyle—from swimming pools to golf courses—is based on water. Finally, although Phoenix is often thought to be one vast retirement community, the median age of its population is below the national median age.

As an experiment in urban growth, Phoenix has two major points of interest: 1) its vision of developing urban villages in the desert, and 2) its freeway funding revolt despite congestion and a relative shortage of freeway lane-miles.

Before the 1970s: A Postwar City

Phoenix was first settled near the present intersection of Washington Street and 24th Street. In 1867, founder Jack Swilling formed a canal company and diverted irrigation water from the Salt River, which permitted the development of farming. Phoenix incorporated as a city in 1881, with a population of 1,700. Recurrent droughts hampered agricultural production. Building the Theodore Roosevelt Dam on the Salt River in 1911 stabilized the delivery of water, and citrus and cotton farming flourished. Then the Air Force arrived and, to keep Phoenix growing, air conditioning. Phoenix, according to one capsule history,

> exists among the cactuses of the Arizona desert for three reasons, all of them man-made. The Salt River was dammed, . . . providing ample water for which pre-historic Hohokam Indians had thoughtfully built canals. The air force liked the clear skies for training pilots in the second world war. And then there came the air-conditioner. . . . In the 1950's Phoenix was a farming and flying town of 100,000 on 17 square miles; to keep cool in summer people slept with wet burlap on their walls.[3]

The area's explosive growth began during World War II, as defense industries followed the military airfields. After the war, people and several large industrial companies flocked to the area for its warm and sunny climate, open spaces, employment opportunities, and cheap land. Because the land in agriculture was available in large contiguous parcels, it lent itself to assembly for development. Relatively good transportation access, the lack of restrictions on development, and low land prices all promoted growth.

Cotton and barley fields and dairy farms quickly gave way. The Valley's first master-planned community, Maryvale, was the brainchild of developer John F. Long who had started mass-producing homes in 1947. Construction on Maryvale began in 1954, and the transformation was quick: "One day, 160 acres stand in barley. Then, 40 days later, a complete subdivision with paved streets and sidewalks and 550 ready-to-move-into homes cover the 160 acres."[4]

Developers began marketing the inexpensive desert real estate and land sales skyrocketed. Individuals bought small single-family lots, intending to either resell them at a profit or develop them in five to 15 years. Other development companies seized the opportunity provided by the increasing demand for new housing to build vast suburban developments, which soon became the model for development throughout the metropolitan area.

Phoenix is a city shaped by transportation—first by the railroads and then by the car. The Maricopa and Phoenix Railroad provided the region's first rail service in 1887, connecting the Valley to the Southern Pacific Railroad at Maricopa. In 1895, Phoenix was connected to the northern part of the state by the Santa Fe-Prescott and Phoenix Railroad.

A main line to the Southern Pacific Railroad was completed in 1926. It connected Yuma, Tucson, and Phoenix. In 1927, regularly scheduled passenger air service began with triweekly service from Los Angeles through Phoenix. These events opened the door for

winter tourism, and Phoenix became a major tourist attraction during the 1920s.

The original survey for the urban area divided Phoenix into sections of one square mile. Property owners were required to dedicate land along the section lines for the road system. These section-line roads have served as an arterial street system. As the urban area grew, the one-mile sections were subdivided into half-mile and quarter-mile sections, and other streets were constructed to provide local traffic services.

The shape of the metropolitan area was influenced strongly by this grid pattern of major streets. By providing almost equal access in every direction, it encouraged low-density growth in all directions. The street grid may have been the most important factor promoting a dispersed pattern of growth. Travel is also naturally constrained by what locals call the hourglass effect, the narrowing of developable land at the middle of the region (near Sky Harbor Airport) because of mountain ranges and Indian reservations.

A serious effort to build major roads and highways for automobile travel did not begin until after World War I. In 1919, a bond issue laid the foundation for a countywide system of highways. The first long-range transportation plan for the metropolitan area was the Street and Highway Plan adopted in 1960. Until the mid-1980s, it was the region's basic document for freeway and highway system planning.

The first streetcar line, which was horse drawn, ran for 2.5 miles along Washington Street. It was built in 1887, and, in 1893, it was replaced by an electric system and a line along Central Street was added. Between 1893 and the early 1970s, several private companies provided bus transportation in the region.

Current Conditions and Recent Initiatives: Growth Management, Water Concerns, and Transportation Funding

Phoenix Compared with Other Metropolitan Regions

In 1990, Phoenix ranked as the third least densely populated large urbanized area in the United States, behind Atlanta and Houston. As in both of these sprawling areas, however, the relative increase in Phoenix's urbanized area from 1980 to 1990 was less than the population growth, producing a net increase in density. The rate of urbanization in these years was ten square miles a year, or 17.5 acres a day—

Figure 6-1

Population of Phoenix, 1970–1995

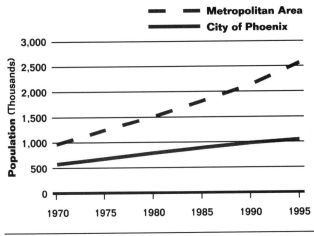

Sources: Decennial Census; and Maricopa Association of Governments.

not quite the one acre an hour popularized by the media, but close.

People tend to think of Phoenix as an auto-oriented community, although in relative terms it ranks about average—18th among the largest metropolitan areas in the use of private automobiles for commute trips (89 percent) in 1990.[5] In number of miles driven daily per resident, Phoenix was also fairly typical among large urban areas—ranking 17th out of 35, as reported in Chapter 1.

A major travel difference between Phoenix and other areas has been the former's lack of an extensive freeway network. In 1990, Phoenix had only 0.3 miles of freeway per capita, compared with 0.52 in Portland and 0.84 in Dallas. This explains why, in 1990, a smaller share of travel occurred on freeways in Phoenix than in Chicago or New York. For example, under 18 percent of vehicle-miles traveled in Phoenix are on freeways, compared with 52 percent in San Diego, 40 percent in Houston, and 28 percent in Philadelphia. However, an aggressive freeway development program that began in the 1980s has resulted in a 10 percent decline in roadway congestion, as measured by the Texas Transportation Institute in its index (see Chapter 1).

The region's transit system is not extensive either. Annual transit trips average only 16 per resident, ranking the city fifth from the bottom among large urbanized areas—ahead only of San Bernardo/Riverside, Tampa/St. Petersburg, Kansas City, and Fort Lauderdale/Hollywood. Improvements in transit services also were made during the 1980s and, while the share of work trips by transit (only 2 percent) remained constant, the number of commuters using transit increased by 55 percent between 1980 and 1990. Total annual bus ridership jumped by 131 percent during the same period, and gained another 10 percent between 1990 and 1994. While the level of transit ridership in Phoenix is low, at least it has positive momentum, unlike transit in many other large urban areas.

Development Market Trends

Population. Metropolitan Phoenix accounts for nearly 60 percent of Arizona's population. Comprising Maricopa County, the metropolitan area contains 23 incorporated cities and towns and numerous developed unincorporated areas within its 9,225 square miles. Maricopa County had a population of 2.5 million at the end of 1995, and has experienced a 3.8 percent annual growth rate in the 1990s. Net migration has accounted for two-thirds of the growth. The city of Phoenix accounts for a little less than half the region's population; in 1994, Mesa contained almost 319,000 people and Tempe and Scottsdale each had slightly more than 150,000 people. Phoenix is one of the largest and one of the least densely populated metropolitan areas of the world. The population is projected to grow to 2.7 million by 2000 and 3.4 million by 2010.

Phoenix has kept up a strong pace of growth since the war. Between 1950 and 1970, the region's population tripled, making it one of the most rapidly expanding metropolitan areas in the United States. During this time, several cities emerged or developed from small farming communities, including Scottsdale and Paradise Valley to the northeast and Tempe and Mesa to the east.

By 1970, the metropolitan area held just under 1 million people. Among its 18 incorporated cities and towns were some of the fastest-growing communities in the country. For example, with a 576 percent increase in population between 1960 and 1970, Scottsdale was the fastest-growing city in the United States. In the same period, Tempe experienced a 155 percent increase in population, and Mesa grew by 86 percent.

The Economy. The Phoenix metropolitan area has five distinct regions: central, northeast, southeast, northwest, and southwest. The central area contains the area's major employment centers, including downtown Phoenix, which has recently undergone a revitalization beginning with the opening of the America West Arena in 1992. The arena is the home of the Phoenix Suns professional basketball team, and starting in fall 1996 will house a National Hockey League franchise. Another major sports center, the

$284 million Bank One Ball Park, is under construction with completion scheduled for 1998. When completed, the 50,000-seat stadium will be the home of a new Major League baseball team, the Arizona Diamondbacks.

The Greater Phoenix Economic Council and the Arizona Economic Council have been active participants with local governments in creating economic development initiatives to attract business to the area. Local and state incentives typically include tax abatements, utility concessions, and relocation funds. These efforts have been most successful in attracting warehouse and distribution development to the southwest region, an area with an abundance of inexpensive land and good access to I-10.

Manufacturing is a leading employer, and the electronics industry is a strong component of this. Phoenix ranks third among electronics production centers in the country. Tourism is also important. Resorts and first-class hotels are numerous. More than 45 percent of total employment is in the retail trade and service sectors. The state and county governments provide more than 30,000 jobs. Phoenix also is the national headquarters for a number of large corporations, including Phelps Dodge, Dial, Circle K, Aztar, U-Haul, and America West Airlines.

Between 1992 and 1994, Phoenix added 118,000 new jobs and had the fifth highest rate of job growth in the nation. The 1.2 million workers in Maricopa County's civilian labor force in 1994 represent a region-wide growth of 74 percent in the labor force since 1980. The city's labor force grew 40 percent in this period, and stands at 600,000.

The region is attracting California industries and firms that are disenchanted with high taxes and stringent environmental regulations. In 1995, for example, the Walter Kieckhefer Company moved from Novato, California, and began construction on a facility that will eventually total more than 700,000 sqare feet. The company told the press that Kieckhefer "just feels [Phoenix] is a more positive business environment [than California]. The government is not trying to put you out of business."[6]

The Japanese industrial giant Sumitomo recently chose Phoenix over Portland, Oregon, for a $400 million semiconductor plant. Phoenix officials think that the plant will attract new development to the Valley and be the key to the city's future vitality in high-tech industries. A package of tax incentives helped Sumitomo choose Phoenix, incentives that had been questioned by the public in Oregon.[7] The proposed site for the plant is the master-planned community of Desert Ridge, in the far north of Phoenix.

Many other manufacturing firms have moved to Phoenix and major employers in the region are expanding their operations and creating thousands of new jobs. These firms include Intel, Motorola (which has 20,000 employees in Phoenix), SGS-Thompson Microchip Technologies, and TRW Safety Systems. Intel's $1.2 billion Fab-12 facility in Chandler is the largest expansion and currently the largest nongovernment construction job in the country. It will create an estimated 4,000 to 5,000 new jobs. In addition, America West Airlines emerged from bankruptcy in 1994, solidifying Phoenix's position as a major transportation hub and preserving jobs.

Downtown. The central business district is bounded by residential neighborhoods of varying ages and conditions, an older commercial district to the south, and a warehouse area that extends south along Central Avenue to the Salt River. As the trade and government center for fewer than 200,000 people, downtown before the war never amounted to much. And downtown did not expand or become notable as Phoenix grew after the war. Two other areas in the core—uptown, which is north along Central Avenue, and the area around 24th and Camelback near the famous Biltmore Hotel—offered better parking and less contact with such urban realities as homeless people. These areas became economically stronger than downtown and attracted development.

Downtown was further weakened by the movement of residents and businesses into the suburbs. In the 1960s, in an attempt to concentrate more activity downtown, the city constructed the Civic Plaza, and new city and county office buildings were built nearby. In 1979, the Downtown Redevelopment Area was established to clear urban blight and encourage development, mostly by means of public sector land assembly powers. The Renaissance Tower, Arizona Center, and the America West Arena are among the projects that were completed under the program.

The city entered into a series of public/private partnerships in the 1980s to encourage private investment in downtown. Using a combination of federal funds and other financial incentives, the partnerships stimulated private developers to build office, retail, housing, entertainment, and recreational projects. Few residential projects have been developed in the downtown area since the 1980s and these have received substantial financial assistance from the city, usually in the form of subsidized land.

One project, the 72,000-square-foot Mercado at Van Buren and 11th, was a failure and seen by some as a sign of the weakness of the downtown retail market. The retail complex, which was built in 1989 by Chicanos por la Causa and the Symington Company and funded through three union pension funds, failed to attract shoppers into downtown. The investors foreclosed and sold the building to a private group, which runs it as a downtown campus of Arizona State University. The sale did not cover the Symington Company's debt on the development, forcing J. Fife Symington, Arizona's current governor, to declare personal bankruptcy.

The Mercado's problems may have had more to do with inexperienced management and the competition of a nearby development, the Arizona Center, which opened 18 months after the Mercado did. The more successful Arizona Center contains more than 1 million square feet of office space, 150,000 square feet of retail space, and a hotel. Unlike the Mercado, it is easy to access by car and has abundant parking.

Downtown appears to be revitalizing. Restaurant and retail development is underway in the Jackson Street Arts Warehouse district, south of the America West Arena. A Major League baseball stadium financed by a voter-approved sales tax is under construction. Other projects in the works for downtown are a new city hall, a new Republic/Gazette building, and major renovation of the Civic Center and Civic Plaza. There are also plans for a new federal courthouse and a new downtown bus terminal. The parking garages for the city hall and the arena will include street-level retail. This development is being supported by a major streetscape renovation project.

The Housing Market. Phoenix has one of the highest rates of homeownership among U.S. metropolitan areas, largely due to the availability of housing in all price categories. As might be expected, single-family housing predominates. Most experts expect that single-family housing will continue to be developed at very low densities and in relatively noncontiguous patterns.

Most low-income families live in central and southern neighborhoods while high-income families live in northern and eastern areas. Since the late 1980s, the city's housing officials have worked to scatter low-income family housing throughout the city. More than half of the city's federally assisted housing units are in the Central City urban village, where most of the housing dates back to the 1940s and 1950s.

In 1995, the city council instituted an infill housing program to encourage the development of single-family houses on vacant lots within the city. The program offers such incentives as waivers of building permit and review fees up to $1,000, waivers of development impact fees for water and sewers, city participation in the cost of off-site improvements, and the services of an infill development team that can help coordinate and expedite projects

Phoenix is one of the most active housing markets and is expected to be among the top five most active single-family markets in the country over the next several years. Approximately 360 subdivisons

are currently active in the metropolitan area. Single-family development is particularly active in the northeast and southeast regions.

The perception that outsiders have of Phoenix as a maze of retirement communities—when, in fact, Phoenix actually has fewer such developments than Dallas or Houston—is based largely on the renown of one of the first large-scale retirement communities in the country, Sun City. Del Webb began thinking about attracting retirees to the Phoenix area in the 1950s. He opened Sun City 20 miles northwest of downtown Phoenix in 1960. By its completion in 1970, it covered 8,900 acres. In 1978, the Del Webb Corporation built another retirement community, Sun City West, next to the original one.

Both Sun Cities followed the same development model: the golf courses, pools, and recreation centers were operational before housing sales began. From the air, the distinctive circular street pattern stands in clear contrast to the regular Phoenix street grid. In Sun City, a recent *Planning* magazine article noted, "golf carts are the transportation mode of choice."[8]

Other developers began specializing in the retirement market, although not at the Sun City scale. Today's Phoenix is dotted with retirement villages ranging from fairly upscale trailer parks to garden apartments and luxury houses. In 1979, the county adopted a senior citizen overlay ordinance, which allowed Sun City and five other communities to enforce their own deed restrictions when they are more restrictive than the county's land use or other ordinances.

The most stringent controls on land use—and indeed on behavior—in the Phoenix region are those imposed by private communities, particularly seniors-only communities. Citizens willingly accept these rules and restrictions when they buy houses in these communities. At Sun City, for example, private deed restrictions require that at least one member of the household be over 55 and that no member be under 19.

Sun City has ten shopping centers, 46,000 residents, and its own quasi government. The private corporation that maintains the community provides everything from a symphony orchestra to police protection. The homeowners association has the power to assess fees to cover the costs of services. If homeowners fail to pay these fees, the association can put a lien on their homes. Homebuyers must offer the association the right of first refusal when they sell.

Homeowners associations regulate down to fine details what people can and cannot do to their property. The Arizona courts have upheld such regulations. Property owners can be told what they cannot park in their driveways, what paint colors and even brands they must use on their homes, and what home improvements they can make. The now infamous developer Charles Keating banned pornographic books from one of his Phoenix developments.

Joel Garreau calls this phenomenon "shadow government" and argues that its invention has its reasons:

> Homeowners readily obey and encourage shadow government. And indeed, such units are very successful at what they do. They control nuisances and unpleasantness and keep the community swimming pool clean. . . . [S]hadow governments devise new solutions to the new problems that [city government] faces every day. If conventional governments had been doing such a great job, people would not have felt obliged to invent new forms.[9]

This view infers that the shadow governments operated by homeowners associations are what has enabled Phoenix to grow and develop despite the ineptness of elected local governments. Local government officials, however, dispute this view. They doubt that private communities have had much to do with growth or development in the Valley.

Paying for only very localized public facilities and services has another attraction, namely that it is substantially cheaper than supporting the facilities and services needed for citywide growth. Residents of Sun City and similar developments have resisted incorporation and have been responsible for defeating school bond elections. One reason is clear: property taxes in Sun City are currently one-third to one-half those in other Phoenix area communities. Of course, not having school districts accounts for much of the differential, and incorporation would not necessarily increase school taxes.

Sun City's taxes are substantially lower than in comparable communities, and its residents also receive fewer public services. In 1996, the Arizona legislature is expected to consider legislation to authorize the formation of "retirement villages," which would be quasi-political entities responsible for some governmental functions like police services, planning, and zoning.

Other Development Markets. The region's phenomenal population growth has fueled **retail** growth. Since 1990, 2.6 million square feet of regional shopping center space has been constructed. Several large power centers are under construction or planned for 1997, including a 200,000-square-foot center at 48th and Ray Road and a project of similar size planned for the southeast region at I-10 and Ray Road.

The **office** market is nearing equilibrium, having recovered from the recent sales of RTC properties and the overbuilding of the late 1980s. The region's office vacancy rate fell from 12.9 percent in 1994 to 10.3 percent in 1995. Little new construction has occurred

during the last five years and absorption has been steady. Four new speculative projects were announced in 1995, including the 250,000-square-foot Scottsdale Spectrum, which is under construction in the northeast region, and the Camelback Esplanade addition. The Esplanade addition will consist of two 50,000-square-foot office towers and a movie theater complex. A 160,000-square-foot office building is planned near the airport. Office projects planned and under construction will add 900,000 square feet to the market.

1995 marked the third consecutive year of rate and occupancy gains for the region's **hotel** market. Demand in 1996 should continue to be strong due to fundamental market strength, local economic development, tourism initiatives, and major sporting events. Several new projects, especially budget and economy hotels, are being developed.

Strong **industrial** absorption in 1995 produced a sharp decrease in the vacancy rate: it dropped from 12.1 percent to 4 percent. Vacancies are at their lowest level in 20 years. Developers are constructing speculative space to meet strong demand. The areas experiencing the most active industrial development—including Tempe, the southwest region, the central region, and the airport area—have superior transportation.

Maricopa County has designated state enterprise zones in distressed areas of Phoenix, Chandler, Tempe, and the west part of the county. The state's enterprise zone program was established in 1990. It gives tax credits of up to $5,000 per new hire over a three-year period to companies that locate in an enterprise zone and create jobs for economically disadvantaged or dislocated workers. Small manufacturing businesses owned by women or minorities also can qualify for an 80 percent reduction in property taxes. Businesses must invest $2 million in the enterprise zone to qualify for the tax reduction. Thus far, the program has had little success in attracting businesses. The current provisions expire in July 1996, and are likely to be extended or modified by the legislature.

In September 1995, the U.S. Department of Housing and Urban Development designated an area in Phoenix as one of 100 enterprise communities (EC) nationwide. The area comprises 20 Census tracts (19.9 square miles) in south Phoenix. The EC designation entitles the city to a $2.94 million grant to promote community-based, sustainable development. The city's plan identifies four program areas: economic development, human development, public safety, and amenities and design.

Growth Controls and Development Fees

Phoenix has traditionally relied on general obligation bonds to pay for capital facilities such as sewer lines, water lines, and major streets, bonds that are repaid by the taxpayers and ratepayers at large. However, in the mid-1980s the competition among projects for bond funds intensified and citizens began to insist that new development pay its own way.

The city of Phoenix's planning efforts in the 1980s were based on the assumption that property values would grow 10 percent each year. But in 1994, Phoenix's net assessed valuations were 10 percent lower than they had been in 1990. The declining tax base prompted the city to adopt a comprehensive development impact fee system in July 1987.

The Development Fee Ordinance requires builders and developers in designated planning areas to pay a proportionate share of the costs of capital facilities needed to serve their projects. The capital facilities included in the system are: equipment repair, fire, libraries, major streets, storm sewers, parks, police, solid waste, storm drainage, wastewater, and water. The projected capital facility needs are derived from formulas based on population and nonresidential space projections from the adopted General Plan.

In the Desert View Tri-Villages area (which is the eastern sector of planning area C/D and the only part of this planning area where development activity is currently taking place), for example, fees are scheduled to reach $10,138 per equivalent single-family house (edu) by 1999. Planners convert all new construction to edus to assess fees. Developers are given credit for specific capital facility and other improvements, depending on the planning area. The fees are collected at the time building permits are issued, and are earmarked for use on specific capital facility categories in the designated planning area as specified by the Infrastructure Financing Plan for that area.

Developers have the option to construct certain improvements themselves, such as parks or water retention basins, in return for impact fee credit. They are given an initial (standard) offset of $2,400 per edu to reflect property taxes and other fees paid, and credited with any facilities that they provide or build that are in the infrastructure program for the area. Credit amounts are expected to change in the next three years as impact fees are increased, but the city does not expect to raise the initial offset. The amount of credit given per edu ranges widely, from zero to roughly $2,600. No developer in recent times, according to city staff members, has paid less than $1,000 per edu after all credits and offsets.

Development fees have been assessed in Phoenix's northern growth areas (north of the CAP Canal) since 1987 and southern growth areas (south of Guadalupe Road) since 1993, and the city council is considering extending the fee system to other growth areas of the city. Slow development in the northern growth areas in the late 1980s means little cash has been collected for the impact fee fund. The city has been waiting for the fund to grow before making any improvements with it, and plans are in progress to begin construction on some projects.

In 1989, the city council set a fee on new development to help fund the acquisition of water resources needed to serve the development. This fee is intended to recover only capital costs related to the purchase and transportation of new water supplies to Phoenix. It will enable the city to fund water purchases without raising water rates, or incurring bonding debt.

In the early 1990s, impact fees were assessed on commercial development in the southwest, between the Salt River and I-10. No building was going on, so there was little opposition from the development community. However, when the development climate improved, opposition to the fees began to surface, and not just from the development community.

The city has never implemented these fees. Some officials report that the failure to do so was due to a finding by the Economic Development Department that the industrial fees might be impeding development. Other city officials insist that the reason was that what little development was occurring was taking place in areas with sufficient existing infrastructure. Thus, the impact fees could not pass the "rational nexus" test and have been shelved until the time that development begins in areas that lack sufficient infrastructure.

Although Phoenix's impact fees were supposed to be reviewed annually, in 1995 they had not been evaluated or changed since they had been implemented seven years previously. In mid-1995, the city proposed phasing in development fee increases in far north Phoenix, to raise the current $4,744 per unit fee to $9,646 or $10,895, depending on the location of the development. The increases, which passed despite intense opposition, are to be used for the construction of water and sewer lines and for some new programs, such as building trails in the desert north of the CAP canal.

The 1995 fee structure adopted by the council raised equipment repair fees effective March 1996 and phased in over four years an increase in street, water, and wastewater fees, beginning in August 1996.

A series of commercial impact fees is in effect in the Desert View area. Developers seeking to build in the area are challenging the fees, saying they make development uneconomic. A grocery store building would pay fees of between $1.5 million and $1.6 million, for example, while a convenience store building would incur roughly $250,000 in fees, based largely on the traffic-generating potential of each project.

Developers and landowners in the area have been encouraged by the city's suspension of fees in the southwest. A developer planning a 5,700-acre master-planned community thinks that the lesson of the southwest experience is that impact fees will stop development. The city's planning director has been heard to admit that there is a conflict between economic development and impact fees. However, a proposal to waive the commercial impact fees in the area in which Sumitomo was thinking of locating was not acted upon because city staff members felt that such a waiver would have negative legal as well as financial impacts.

For Phoenix, the basic issue in the Desert View area is whether the commercial impact fees make retail development more expensive in the city than it would be in neighboring Scottsdale. The generation of sales tax revenue hangs in the balance. The mayor of Phoenix has proposed a moratorium on retail development incentives, and the city of Phoenix and the city of Scottsdale are negotiating a shared sales tax zone for the Desert View area.

Commercial development requires infrastructure and, in the view of the city's planning staff, there is no funding source other than impact fees for the facilities needed. The Maricopa Association of Governments—the region's metropolitan planning organization—recently completed an urban form study that encourages impact fees:

> The equity of new development paying for its directly related infrastructure costs should be reinforced. Leap frog development, if allowed, should be required to pay for the extra cost of extending infrastructure. Similarly, rehabilitation and infill efforts that exceed planned transportation improvements should be assessed accordingly.[10]

Other cities in the region have been slow to adopt any system of development fees. Most make a strong (if arbitrary) distinction between sewer and water hookup fees and impact fees to finance off-site improvements—even though hookup fees also can add up to thousands of dollars per unit. While the Phoenix fee system is based upon infrastructure needs that are defined in the Infrastructure Financing Plan, other municipalities in the region are less specific in their fee designations and are more likely to require capital improvements from developers.

The city of Mesa has established a system of development fees in new growth areas, but only for water and sewer development. Tempe and Scottsdale have no formal impact fee systems, but both impose development fees related to water and sewer hookups. Scottsdale currently charges from $1,015 to $81,190 in sewer fees for commercial construction (the fees varying with the size of the meter); its water fees for commercial construction range from $790 to $62,485. When Scottsdale extends service beyond its corporate limits, it imposes a 125 percent surcharge on these rates. The water hookup fee for single-family homes in Scottsdale ranges from $790 (for a 5/8" meter) to $3,910 (for a 1.5" meter); Scottsdale's water resource fee, adopted in 1987, is $1,000 per single-family dwelling unit.

Water: Will Supplies Support Development?

Phoenix lifestyles suggest an abundance of water. Do Phoenix residents with their plentiful manmade lakes, swimming pools, and golf courses waste water more than do their southwestern neighbors? Will water quality, unreliable precipitation, unresolved water rights claims, and inadequate water supplies really have a limiting effect on future growth? There is much local debate on these questions.

The substantial controversy over whether current levels of consumption can continue is largely related to the process by which Colorado River water is allocated among the states now using it: Colorado, Nevada, Arizona, and California. To some observers, barring changes in the Colorado River allocation process or a severe drought, the Valley of the Sun has no water problem. To other observers, water shortages may lead to severe restrictions on development in the future.

Before World War II, water was supplied through wells, a large reservoir, a system of canals, and subsurface water from the Verde River delivered through 30 miles of above-ground pipe. The Theodore Roosevelt Dam on the Salt River was completed in 1911 and supplied water and electric power to the Valley. A system of multipurpose dams constructed on the Salt River in the 1920s and a network of irrigation canals assured the continuation of the agricultural economy.

The postwar shift to an industrial economy and rapid urban growth led Phoenix to reconsider its water sources. The city dug deeper wells, constructed water treatment plants, and purchased private water companies. It completed the Salt River Project canal system to deliver water to the city.

As the city limits expanded in the 1960s and 1970s, the capacity of water treatment plants had to be increased and more reservoirs had to be built. Phoenix continued to purchase private water companies to assure water service for new growth areas. In 1973, construction began on the Central Arizona Project (CAP) canal system, which was designed to bring water from the Colorado River to the urbanized areas of southern and central Arizona. The project's promise of water encouraged more growth and development in the Valley.

The Salt, Verde, and Colorado rivers supply more than 90 percent of Phoenix's water. The other 10 percent is groundwater.

In the 1980s, the focus of water policy was conservation and completion of the CAP canal. In 1980, the Arizona Groundwater Management Act required cities to establish a permanent conservation program for agricultural, industrial, and municipal water users. The act's purpose was to reduce groundwater use. The city of Phoenix responded with its Long-Range Water Resources Plan in 1985, which it updated in 1987 and 1990.

Based on projections of water supply and demand, the long-range plan established guidelines to assure an adequate supply of water until 2035. A series of demand management and conservation programs has been promulgated. Conservation programs for residential landscaping, industrial uses, and city recreation areas have been put in place. The use of treated effluent from wastewater treatment plants for recharging underground aquifers and irrigation is being considered. Other conservation measures include pricing mechanisms, the installation of water-conserving fixtures in older homes, changes in building codes, regulations requiring businesses to adopt new technologies, and limits on lawn sizes. Most other communities in the region also have adopted water conservation strategies of one sort or another. Several water reclamation plants are under construction to supply large users with nonpotable water.

As a result of these planning efforts, per capita water consumption in the Valley has been dropping since 1975. In fact, subtracting agricultural water use, per capita water consumption in Phoenix is only slightly higher than in Los Angeles, and it is lower than in Las Vegas.

Projects are also underway to augment water resources. The city of Phoenix is obtaining water from the McMullen Valley land purchase, which encompasses14,000 acres of purchased land and 2,000 acres of leased land in western Arizona. The Roosevelt Dam has been heightened to increase reservoir capacity. A new Waddell Dam on the Agua Fria River is planned to store CAP water.

A key issue for Phoenix's growth is the future distribution of Colorado River water. The problem is one of competition between agriculture (especially agriculture in southern California) and urban users more than it is one of competition among urban users. In the states sharing the river's water, between 70 and 85 percent of all water use is for agricultural purposes. Some planners feel that the most important constraint on growth is not the amount of water available, but the ability to build the infrastructure to take the water to newly developing land.

Urban Villages in the Midst of Sprawl

Concerns about the anonymous nature of the region's low-density residential development sparked a vision for the future in the 1985 Phoenix General Plan:

> Great cities satisfy a wide range of aesthetic, economic, and social needs. They give residents a sense of identity. They provide a mix of jobs, a range of housing types and prices. They protect and preserve the natural environment; they offer recreational, cultural and educational opportunities. They allow citizens a chance to participate in shaping the future of the city.[11]

The concept of urban villages emerged as the unifying element of the plan. This concept had been developed by a citizens committee in the late 1970s. The city's 1985 plan established nine villages, nine urban village planning committees, and the urban form for Phoenix. The villages, it was hoped, would satisfy the "psychological need to belong to an identifiable community within a large city." The plan has since been modified, the last time in 1994. Today there are 11 villages.

Like all planning efforts, the urban village plan is an ongoing process. The city has made and will continue to make adjustments in reaction to changing market conditions, new information, specific environmental issues, and implementation difficulties. A planning process that attempts to balance numerous interests and to regulate growth and development encounters many problems. Some have been resolved, others not. Some are still to come.

To begin, the General Plan has seven objectives. The emphasis and priority accorded each have changed over time. They are:

- Balance housing and jobs;
- Concentrate intensity in village cores;
- Promote the individuality of each village;
- Preserve and enhance the quality of life in each village;
- Provide for most needs of residents within the village;
- Direct urban planning through the village planning committees; and
- Balance economic impacts and land use decisions.

The original goal was to help each village grow by balancing employment and housing, which would produce a regional jobs/housing balance. The 1985 plan projected that the nine villages would together have 1.1 million people by 2000, living in just under 500,000 homes, for an average 2.29 people per unit. Each of the villages, however, would develop individually and have its own demographic and housing mix.

Figure 6-2

Transportation Indicators for Phoenix, 1970–1990

	1970	1980	1990	Percent Change	
				1970–1980	1980–1990
Population	967,522	1,509,052	2,122,101	56%	41%
Commuters	365,896	658,854	996,495	80	51
In Private Vehicles	325,190	587,125	890,998	81	52
Driving Alone	287,568	517,967	814,074	80	57
Passengers	37,662	69,158	76,914	84	11
By Transit	4,256	12,870	19,962	202	55
Commute Share					
Transit	1%	2%	2%	68	3
Private Vehicles	89%	89%	89%	0	0
Travel					
Daily Vehicle-Miles Traveled (Thousands)	15,219	18,656	39,654	23	113
Annual Transit Trips (Thousands)	n/a	12,804	29,581	n/a	131
Median Single-Family Lot Value	n/a	$20,000	$30,000	n/a	50

Sources: Decennial Census; *Journey-to-Work Trends in the United States and Its Major Metropolitan Areas, 1960–1990* (Federal Highway Administration, PL-94-012); and ULI residential land price surveys.

The plan established the boundaries for the villages and their higher-density cores, and development goals and policies. Each village was to have between 100,000 and 150,000 people, and its own version of physical, social and urban characteristics. Each core, where employment and most services were to be located, was designed to provide jobs for a majority of the village's residents. The core would also have 50 percent of the village's multifamily housing (above 15 units per acre). The cores could also house some regional functions—educational or cultural facilities, for example.

For each major urban village goal, the plan laid out a number of supportive policies. For example, in order to balance jobs and housing, each village should:

- Provide jobs equal to 45 to 55 percent of the number of residents;
- Favor development proposals that improve the jobs/housing balance;
- Encourage the development of land already zoned and vacant, rather than rezoning land;
- Encourage a 40/60 split between basic and service jobs; and
- Develop regional shopping centers in villages lacking them.

A master plan for each village core would be developed by the village planning committee in conjunction with landowners and homeowners. When approved by the planning commission and city council, village plans would be integrated into the city's General Plan. The village plans would identify the desired mix of uses, separate pedestrian and auto traffic, reduce through traffic as much as practical, and create an integrated urban design.

Very specific policies guided the development of the village cores to ensure that they became the identifiable central focus for their villages. The core was to contain the greatest height and most intense uses "within limits based on village character, land use needs, and transportation system capacity." Building heights were to be negotiable, based on the amenities, mix of uses, and infrastructure improvements offered by developers. The core was to be pedestrian-friendly, with "plazas, common open space, shaded walkways, and the separation of pedestrian and vehicular traffic." And it was to contain a wide variety of land uses, which could include industrial buildings as well as commercial facilities.

Today, while the villages share some common themes, each is designed to have a different mix of land uses and building heights. For example, the Deer Valley Village in northern Phoenix is slated to have high-tech employment and airport industrial businesses while the Maryvale Village is planned as a blue-collar family community with small-lot single-family homes.

In 1989 and 1990, the city sponsored a series of citizens meetings, called the Futures Forum, to discuss and articulate a vision for the future. Some of the forum discussions focused on the strengths and weaknesses of the urban village concept. In 1991, as a result of these discussions, the urban village concept was refined "into a new urban form model for Phoenix," which was adopted by the city council in 1994.[12]

When the city then undertook an analysis of regional growth patterns and trends, it concluded that there were several different labor markets, each of which was, or could be, centered around current and future employment centers. The city designed a series of strategic actions to strengthen existing employment centers and assure that both the inner city and the regional employment centers would capture their fair share of regional employment growth .

Approved by the city council, the new strategic focus on regional labor markets led to a new model of future development for Phoenix focused on regional centers that extended beyond the original village cores (and which might not be in the same place). These regional centers were tied to (sometimes overlapping) commute sheds that contained more than one urban village. For example, the seven central-city villages were in the commute sheds of several regional employment centers, including the airport, the southeastern industrial area, the Camelback core, and Central Avenue. The most recent plan revisions also rethink the role of the downtown, seeing it as a regional center rather than as the core of just another village.

The urban village concept was also changed to reflect market constraints and forces. In particular, it has been recognized that not all the commercial development in an urban village can or will occur in the core, and that certain centers of activity outside the core can be a focus of the urban village, although they may not be employment based.

Critics charge that the new regional employment centers plan diminishes the capacity for achieving a jobs/housing balance and weakens the neighborhood or community focus of the original vision. In contrast, city officials feel that some of the new elements in the plan actually strengthen the plan's capacity to bring about a jobs/housing balance. Moreover, they point out that neighborhoods are a crucial focus of the new plan and that neighborhood groups have consistently used the neighborhood element of the new General Plan to support local conservation efforts.

The revised plan is substantially less anticar than the original version. It does not assume that people will be willing to walk or use transit for trips between elements in the villages—from home to a job in the core, for example—but it does put far more emphasis on pedestrian movement within elements, such as the core or residential neighborhoods. City planners think that the new plan's recognition of the relationship between neighborhoods and regional employment centers supports transit use. In fact, the street design portion of the General Plan now includes transit elements.

Has the urban village model helped curb urban sprawl? Some critics see the constant plan revisions as indicating the unworkability of the concept, while others see them as indicating the lack of political will to make the hard decisions that are needed for the concept to work. Proponents of urban villages often talk about the continuing evolution of the concept as evidence of its strength and flexibility. But the overall density of the city has remained low. It is not clear that the plan has made a major difference in the way the city has developed.

Many people in Phoenix think that the urban village plan has had little or no impact on development. "Urban village" is a term that is not used much by anyone. A six-part series in December 1995 in the *Arizona Tribune* on the region's rapid growth never mentioned urban villages, although city planners were interviewed for the series. The term "sprawl" was used more than four dozen times.

The critics of the urban village concept in action are shooting at a moving target—one that has been revised and updated several times in the last decade. That said, there have been eight major criticisms, as follows:

Unnatural communities. The original nine villages were very large. Their boundaries were drawn for political and practical reasons—and not because they encompassed any kind of natural social or human communities. The size and structure of each village had little to do with the real-world catchment areas of commercial or industrial establishments, or with how businesses interact.

Shopping malls as the focus of communities. Critics question whether malls and shopping centers can be a focus of their communities in the same way that traditional downtowns were. According to the *Arizona Republic*: A bunch of low-density subdivisions around a regional mall does not constitute an urban village, but is "just sprawl with a fancy name." Calling an area an urban village does not make it one, the paper editorialized.[13]

City staff, on the other hand, note that while shopping malls have not generally worked in the way that old-fashioned downtowns did, they do provide a regional urban focus. They say that the Paradise Valley Mall at Cactus and Tatum—an area which the *Arizona Republic* editorial said does not deserve to be called an urban village—provides a retail/entertainment focus for its village, even if it is not a community focus in a social or village sense.

In 1992, the city's mayor wondered if locating shopping malls in the middle of urban villages hurt the city's sales tax base. He noted that malls elsewhere in the region were located on the edges of their cities and thus attracted shoppers from Phoenix. In 1992, annual sales tax collections had declined for the first time—by 2 percent. However, this did not seem to be a problem unique to the city. All the cities in the Valley experienced a drop in retail sales performance in 1992. While sales at Phoenix department stores were lower than department store sales in adjacent communities, they accounted for only 6 percent of total sales. General retail sales per capita in Phoenix, which accounted for 50 percent of total sales, were higher than average general retail sales throughout the Valley.

Deviations to accommodate development. Critics charge that the city has allowed substantial deviation from the concept in most of the nine village areas, largely to accommodate private development and the growth of shopping malls. A 1992 editorial in the *Arizona Republic* charged: "After the city invented the urban village concept in 1979 to control urban sprawl, developers spent the next decade inventing ways to circumvent it. Now Phoenix has slashed its planning staff so drastically that the concept might never recover, even without the developers."[14]

Ron Short was the planning director then. When the planning department's budget was slashed, he quit. Short said that the city lacked the staff to implement the urban village concept.[15]

Undistinguished villages. Another major criticism is that city efforts to create individual characters for the 11 villages—a difficult task—have been minimal. In particular, the failure to develop strong and unique design standards for each village has been criticized.

The village planning committee process. The way that the village planning committees work has been criticized. Committees hear requests for zoning changes and variances as well as general plan changes, and they address parks, street improvements, public facilities, police and fire service, and other development issues. But some critics think that the committees are not representative of the people whose needs they are supposed to be addressing. A 1992 study by Arizona State University's Morrison Institute found that the typical committee did not really know its neighborhood or encourage public participation.

A related criticism is that the city may not really listen to the village committees or give them any power over their environment. In 1990, the process of updating individual village plans made a number of village committees unhappy. The Camelback East Village Committee's revised plan contained regulatory sanctions, which the city planning commission removed. The Paradise Valley Committee complained that what the city planners wanted was "a meaningless plan" lacking regulatory powers to enforce guidelines.[16]

High-density and mixed-use development in practice. Application of the concept of a mix of land uses and higher densities in neighborhoods has sparked some antigrowth sentiment, which some members of the development community think threatens the urban village concept.

In the Camelback East Village, a multistory commercial building was allowed to back up against existing single-family homes. Because it was in the core, there was no requirement for a setback from existing homes. This was seen by local residents as commercial development eating into a stable neighborhood, and as a sign that the city had lost control of the core. They blamed "chaotic city planning influenced by greedy developers and the whims of city administrators."[17] The city planners, on the other hand, saw this as a small flaw in the original process and attempted to remedy this and related problems as part of the ongoing revisions to the plan.

However, the idea of locating commercial land uses so close to existing or planned residential neighborhoods has continued to generate controversy. As the December 1995 *Arizona Tribune* series on sprawl noted:

> For every resident demanding amenities in a neighborhood, such as shops and restaurants within walking distance, there is another resident refusing to have them next to his home. Concepts like "mixed use" and residential "diversity" sound good in theory, but try selling theory at a zoning hearing packed with homeowners.[18]

In the still largely rural northern planning area of the city, there is erupting a similar controversy over mixing land uses. The Desert Ridge area was designated in the original urban village plan for a mix of high- and low-density residential, industrial, and commercial uses. Desert Ridge is now under consideration by the city as a regional employment core under the 1994 revisions to the urban village plan.

In the meantime, proposals from two companies—Mayo and Sumitomo—to build facilities in Phoenix have prompted the city to move rapidly to implement zoning that would permit them to locate in Desert Ridge. The land that Mayo has picked had originally been designated residential; that chosen by Sumitomo had been designated industrial.

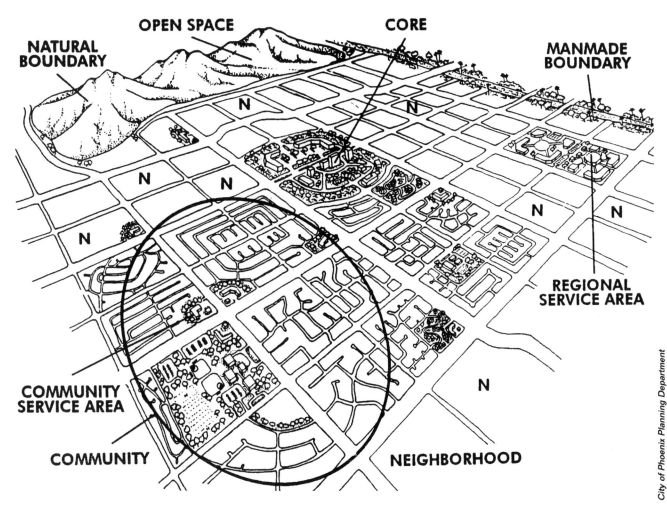

NATURAL BOUNDARY

OPEN SPACE

CORE

MANMADE BOUNDARY

N N N N N N N N N N N

REGIONAL SERVICE AREA

COMMUNITY SERVICE AREA

COMMUNITY

NEIGHBORHOOD

City of Phoenix Planning Department

In the 1994 update to its General Plan, the city of Phoenix refined the urban village concept into a new model of urban form that seeks to balance jobs and housing on a more regional basis. Shown here is a rendering of a proto-typical new urban village.

The proposed new Desert Ridge regional employment center will be less than a half mile from low-density suburban developments, whose residents are strongly opposed to its implementation. In fact, homeowners are pressuring the city to rezone the Sumitomo land as residential. While a number of issues are embedded in this controversy, clearly the mixing of relatively high-density industrial development and low-density housing has not been welcomed by many of the people who live in the affected area.

Lack of regional support for the concept. Urban villages as a concept or a planning reality seem not ever to have been represented in any way in regional highway or transit plans. Moreover, other governments in the county have shown little support for the basic concept.

In 1994, the Maricopa Association of Governments (MAG) conducted a study of alternative congestion management systems. The study compared the urban village model of development (which it called "con-centrated activity centers") with a "dispersed/balanced" model. The urban village pattern was found to result in more congestion and more air pollution, even though it produced fewer vehicle-miles traveled and greater transit usage.[19]

A 1995 MAG study concluded that congestion would increase if the region developed along the lines of the urban village model.[20] Transportation facilities would be overloaded and carbon monoxide emissions and concentrations would increase. Also, infrastructure costs would be higher and the quality of life would not necessarily be improved. In fact, on many study criteria, the results under the urban village model of urban growth were worse than under the other models analyzed. Urban villages fared better only against the base case—that is, the do-nothing alternative.

Haziness of the urban village concept. It is clear that very few people mean the same thing when they describe an urban village. City staff members take great pains to show how development that differs considerably from what was originally envisioned somehow supports the concept. They insist, in effect, that the exceptions prove the rule. For others, the exceptions prove that the rule has been bent beyond recognition. A prominent Phoenix developer calls a residential community that he has built an "urban village," although it lacks a core or any commercial facilities. Other people who are knowledgeable about planning issues consider the concept to be fairly hazy. They say that only city staff members have any idea of what it really means.

Transportation: The Financing Revolt

Freeways. Given that Phoenix is a really a post–World War II city, one might imagine that its urban form was shaped by a network of freeways. In fact, Phoenix did not begin large-scale freeway construction until the 1980s and large portions of the region are not served at all by freeways.

The 1960 Street and Highway Plan called for the development of 140 miles of freeways and expressways to be integrated with a 375-mile major street program by 1980. The planned system still holds up

to scrutiny today, but by 1980 little progress had been made in implementing it. Only 40 miles of freeways had been built by 1977.

In October 1985, in response to traffic congestion around the Valley, Maricopa County voters passed Proposition 300 introducing an excise tax to be dedicated mostly to freeway construction, with a small percentage going to transit. The additional sales tax was set to equal 10 percent of the state sales tax rate, or 0.5 percent of most taxable items. It went into effect on January 1, 1986 and was set to expire on December 31, 2005. In addition, Phoenix had access to HURF money (the Arizona Highway User Revenue Fund), roughly 15 percent of which may be used throughout the state for limited access highways. Phoenix receives about 75 percent of HURF funds.

In 1986, MAG developed a freeway/expressway plan calling for a network of 320 miles of freeways and expressways by 2005. Roughly 228 miles were to be financed through bond sales supported by revenue from the excise tax. An additional 85 miles were to be (or had been) financed with a combination of HURF money and federal funding.

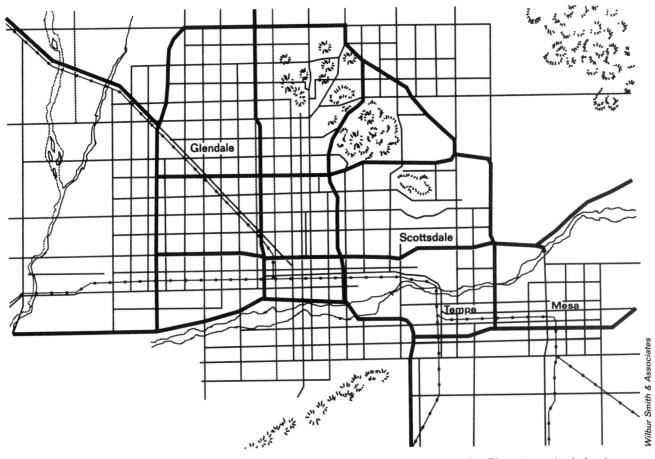

Until the mid-1980s, the long-range Street and Highway Plan adopted in 1960 was the Phoenix region's basic document for freeway and highway system planning.

Freeways were identified as either Proposition 300 freeways or non–Proposition 300 freeways. Between 1985 and 1989, 13 new miles of both types of freeway were opened, including portions of the Papago freeway (in the middle of Phoenix), the Squaw Peak freeway (running north from the airport to the planned Pima freeway in the far north), and the Agua Fria freeway (running east/west in the northern part of the city). In 1990, 28 miles of freeway were in construction and 55 additional miles were programmed for completion by FY 1994.

However, the 1986 MAG plan quickly encountered financial problems, as costs rose and tax revenues fell. Forecasts had suggested that excise tax revenues in FY 1991 would be $117 million; actual collections were $113 million.[21] At the same time, the cost of acquiring rights-of-way increased dramatically. Land acquisition costs for the entire plan had been estimated at between $700 million and $900 million. By 1990, more than $1 billion had been spent, and another $1 billion was budgeted for additional purchases.

The city of Phoenix responded to the decline in funds available regionwide for transportation by issuing bonds to pay for roughly five miles of the Squaw Peak freeway and the state paid for an additional two-mile segment. In March 1989, the Valley's voters were asked to increase the excise tax to 1 percent of taxable sales, in order to alleviate the freeway plan's financial problems and finance a rapid-rail transit system. The voters refused, and thus left the region without the resources needed to implement the recommended long-range plans.[22]

In 1990, MAG concluded that the construction program would need to be extended to 2015 and that the cost for Proposition 300 freeways alone would be $6.3 billion, which was $2.9 billion more than available funding.[23] The funds required for the non–Proposition 300 freeways would be $3.1 billion above initial estimates through FY 1996. MAG reported that Proposition 300 freeways could be completed using new revenues—for example, the extension of the half-cent sales tax beyond 2006, an additional sales tax, higher gasoline taxes, or impact fees. Or, the agency reported, completion dates could be extended beyond 2005, with the possibility that some elements would be delayed indefinitely.[24]

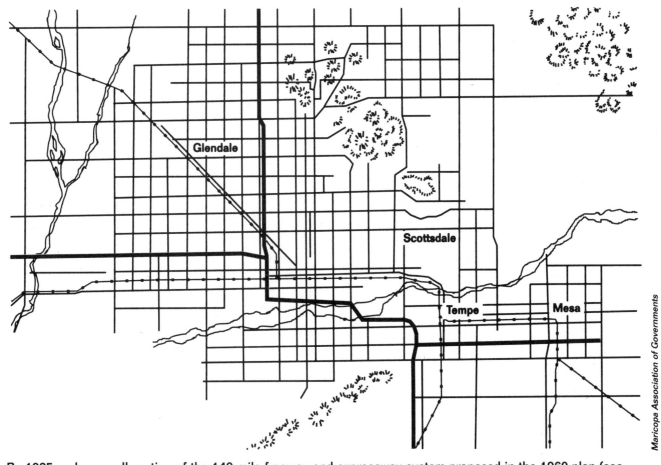

By 1985, only a small portion of the 140-mile freeway and expressway system proposed in the 1960 plan (see facing map) had been implemented. Voters passed Proposition 300 in 1985 to fund more freeway construction.

Maricopa Association of Governments

In 1992, MAG created a fiscal analysis unit to provide yearly updates on the status of freeway construction and financing. It was charged with reporting "to the public regarding the status of the MAG program and the fiscal implications of changes in the MAG program scope, schedule, priorities, costs, and revenues."[25]

Even though financial difficulties were looming, the freeways that were built from 1986 to 1995 had dramatic short-term impacts. Between 1988 and 1992, Phoenix increased its freeway capacity by 1.86 percent per year—substantially more than in any other urbanized area in the United States. San Diego increased its freeway capacity during this period by 0.16 percent per year, Miami by 0.50 percent, and Houston by 0.88 percent. The new freeways clearly had an impact on congestion. Between 1982 and 1992, the region was still growing rapidly, but congestion in Phoenix dropped by 6 percent, making it one of only two U.S. urbanized areas to reduce traffic congestion in this period. While total travel (VMT) went up almost 5.5 percent per year in that decade, Phoenix went from being the fourth most congested U.S. urban area to being the 21st.

The explosion of freeways was thought by many to have accelerated the speed of land development and growth. The growth in the east along the Superstition freeway and north along I-17 would have eventually occurred without the network of highways, but their presence certainly expedited the development process. At the same time, some of the fastest growing areas in the Valley, especially north of Phoenix and Scottsdale, are nowhere close to any of the new freeways. Most observers expect that areas south of the CAP canal and east of I-17 will be built out long before the proposed extension of the Squaw Peak freeway reaches them. Similarly, growth in the east will occur at some distance from existing or proposed freeways.

In November 1994, the freeway funding issue went back to the voters. An extension of the excise tax to 2015 and an additional 0.5 percent excise tax (also to expire in 2015) would fund freeway construction and transit—divided more evenly between roads and transit than in the original and 1989 proposals. By this time, the tax revenue bonds that had been issued to finance the freeways had almost reached the state limit for all the funds due to be collected through 2005. Extension of the tax would permit the region to issue new bonds. However, voters again rejected the funding proposition, by 46 percent to 54 percent.

Proponents of the 1994 measure argued that population growth in the next decade would increase VMT by roughly one-third. In the absence of new freeways, average travel speeds at rush hour would drop by 10 percent. Opponents successfully argued that the planned system was adequate and could be completed under the current tax financing structure. Many people criticized the regional plan for constructing the freeway network in "bits and pieces," like an overpass in Mesa at McKellips Road—on which legislators were photographed—that connected nothing. Two maps were distributed: the regional system proposed to voters in 1985 and the disconnected series of freeway segments that had actually been constructed by 1994.

The failure of the 1994 vote had many causes. Many voters thought that new freeway construction was unnecessary, and that taxpayers should not be burdened with additional taxation for it. Some opposition, especially in Sun City and Sun City West, was opposition to any new taxes. Only those communities in the Valley that would have gotten additional freeways voted for the additional tax, indicating that many voters may have failed to consider the issue from a regional perspective.

The feeling that freeways only encourage growth and sprawl contributed to the failure of both the 1994 and the 1989 propositions. Having come from more congested regions, many new residents of Phoenix—who make up a high share of all residents—tend to see Phoenix as relatively free of congestion, and they see no reason to vote additional taxes for something that is not even a problem.

A local reporter identified what was probably the single most important reason for the failure of the funding propositions: "One word: Trust. The lack of it sank Proposition 400."[26] Government officials had been unable to build a freeway system with funds thought to have been adequate at the time. The extremely large difference between the system's planned and actual costs seemed to indicate poor planning and management. Why should the voters support this?

In December 1994, Governor Symington presented another plan for financing the completion of the MAG freeway program. This included higher-than-forecast sales tax revenues, earmarking more regional federal funds for freeways, and budget savings—through eliminating some freeway corridors, reducing the number of lanes planned, and cutting lighting and landscaping costs. The governor also proposed raising the state's limit on the amount of future tax revenues that could be pledged as bond collateral.

The governor's plan, which was incorporated into the MAG regional transportation plan, also called for accelerating the completion of the so-called outer loop—the Agua Fria freeway has been accelerated one to five years while the east leg of the Pima freeway has been accelerated one to four years—and several other sections of freeway. Finally, the governor's plan urged cities to fund their own freeway segments.

While the business community was solidly behind Proposition 400, many developers in the region are not particularly worried about the segments of the freeway that will be deferred by its defeat. SunCorp, a firm with land holdings that are substantially affected by the freeway delay, feels that, when development reaches the periphery, residents will be more willing to pay for freeway construction and localities will then have the necessary resources.

Transit. In 1971, the city of Phoenix assumed responsibility for the transit services provided by private companies. Other cities in the area were also operating transit systems. In 1985, the passage of Proposition 300 established the Regional Public Transportation Authority (RPTA) as a means of bringing some order to the multiple services and obtaining funding for transit operations and, possibly, the construction of a light-rail system. The RPTA was charged with planning and operating transit in Maricopa County. Funding was provided in part by the excise tax approved in Proposition 300.

In 1995, RPTA's transit operations budget was $68 million. Roughly 30 percent was from riders; about 21 percent from the state's Local Transportation Assistance Fund, which is generated from lottery revenues dedicated to transit; about 10 percent from the added sales tax; about 5 percent from the federal government; and roughly 32 percent from the seven cities in the region receiving or providing transit service—Phoenix, Mesa, Scottsdale, Tempe, Glendale, Chandler, and Peoria.

The RPTA serves as the umbrella organization for most of the region's transit services. The transit system it administers consists of four independent bus systems (in Phoenix, Mesa, Tempe, and Scottsdale) and eight dial-a-ride services. While each bus system and dial-a-ride service is funded primarily by its local jurisdiction, the RPTA has helped institute regional fare and transfer policies, passenger information, and customer service. Currently, 53 local bus routes and 19 express routes serve the metropolitan area.

The RPTA was affected by the defeat of Proposition 400. The agency intended to use the funding that this would have provided for transit to double bus service—largely by adding late night and Sunday services, expanding express bus services, and replacing diesel vehicles with gas- and electric-powered buses—and create a 24-hour, seven-day dial-a-ride service.

The RPTA intended to use the added funding also to help pay operational costs, which had increased faster than had been projected and were being funded by declining sources of revenue. Contract costs for bus service were increasing faster than were sales tax revenues, while federal funds for operating assistance were declining, and revenues from the Local Transportation Assistance Fund were lower than had been projected. Without a dedicated funding source, the RPTA is preparing for difficult times ahead by streamlining services, increasing fares 25 percent, and delaying plans for the expansion of bus and dial-a-ride systems. Resources for long-range transit improvements are lacking.

Significantly underfunded, the regional transit system provides less than half the service available in other western cities. While Phoenix is the seventh largest U.S. city, its transit system ranks 42nd in size. It is the only major metropolitan area without a Sunday bus service. As a result, transit use is relatively low. In 1990, Phoenix residents took only 16 transit trips per capita, compared with 29 in San Diego, 31 in Houston, and 32 in San Jose. However, according to the RPTA, because of recent streamlining, the Phoenix transit system has become the fourth most cost-effective transit system in the United States. By eliminating lightly-used suburban, Sunday, and other low-density routes, the system has significantly increased passenger loading per operational hour.

Before the 1994 election, commuter rail and a light-rail streetcar circulator for central Phoenix were under consideration. In 1993, an Arizona Department of Transportation study offered one option: a 75-mile commuter-rail system consisting of a 20-mile central spine running northwest/southeast through the city with four branches going northwest, west, southeast, and south. However, the lack of voter enthusiasm for supporting transit and freeways makes the proposed Phoenix circulator and other rail systems seem, at best, long-range ideas.

Transportation Control Measures. Air pollution in the Phoenix region is related mostly to automobile use. Since 1989, the region has implemented three programs to reduce carbon monoxide emissions to comply with federal clean air standards. The county's mandatory oxygenated fuels program requires gasoline retailers to sell only oxygenated fuels—gasoline blended with an octane enhancer (methyl tertiary butyl ether) or with alcohol—from October 1 to March 31.

An employer-based travel reduction program requires public-sector and private employers in Maricopa County with 100 or more full-time equivalent employees to designate a transportation coordinator, determine the commuting patterns of the employees, provide employees with information on alternative commute modes, and develop annual travel reduction plans. A regional rideshare program operated for MAG by the RPTA works with businesses and local jurisdictions to inform commuters about alternatives to the single-occupant vehicle.

The third program, a state vehicle inspection and maintenance program, identifies vehicles that do not meet emissions standards and requires that they be fixed.

In its 1995 report, MAG indicates how much carbon monoxide reduction each of these programs has achieved: the oxygenated fuels program has reduced carbon monoxide by 16.2 percent, the vehicle inspection program has reduced it by 4 percent, and the travel reduction program has reduced carbon monoxide by 1.8 percent.

Outlook: Making Changes

Phoenix has addressed the relationship between development and transportation by attempting to create an urban form—the urban village—that allows people to live close to where they work and shop and play. At the same time, the region has been forced to plan for the transportation improvements needed by the growing community within the financial limits imposed by taxpayer revolts.

The urban village concept is a thoughtful, ambitious, and sophisticated effort to come to grips with low-density growth. But, as Roy Drachman, an Arizona developer, points out: "Enthusiasm for the concept has waned considerably, even among the planners and certainly among the elected officials, who would not have allowed cuts in the planning staff if they thought that voters strongly favored the urban village idea." The city is still committed to the idea, even if it means different things to different people. A few planners and citizens are firmly committed to the original model of a number of individual villages, each offering all its residents opportunities to live, work, shop, and play entirely within it—and often within walking distance.

Many others see the concept as having evolved into a different model that does not try to force all communities into a single type of urban form. They recognize that no one model of urban growth and development will work in all circumstances, and also that the economic goals that the plan established for individual villages may not be feasible. In particular, the originally-designed urban villages are too small to provide a sound employment base for their residents.

There is a growing feeling that the city's effort to implement the urban village concept can work and make a difference if planning is tailored to each community and neighborhood (allowing one community's plan to differ greatly from another's); if the focus of the jobs/housing balance is widened to encompass areas larger than single villages; and if citizen involvement is facilitated.

Earlier planning had ignored two important considerations. First, people want different things from their communities, so one urban village model cannot satisfy everyone's needs. Second, it is not possible to create a jobs/housing balance in a small area. Employment planning must conform to natural commute sheds.

The city's current planning approach recognizes a new goal of balancing jobs and homes across the city. Economic strategies are developed not for each village but for entire large commute sheds. The latest revision of the plan identifies three major employment growth corridors—the east Pima freeway corridor (where Sumitomo is located), the far north I-17 corridor, and the west I-10 corridor—and the city has begun to articulate new development strategies for each one.

Also, the city's planning staff is currently evaluating the role of the village planning committees. When budget cuts were made in 1992, planning committee staffs were sharply reduced, all but negating the contribution of the committees as a primary means for public participation in planning decisions. The planning staff hopes to document the effects of the staff cuts on public participation and to develop a new planning process.

In the latest revision of the plan, the urban village concept is treated more as a vision than as a requirement:

> The model will be implemented through the policies of the General Plan and the mechanisms for implementing those policies. The model represents a desired end state. However, because of the dynamics of urban development, the desired end state will likely never be achieved by the City as a whole. It may be achieved within small portions of the City, particularly newly developed areas. For the more urbanized parts of the City, this end state provides a model for which portions may be implemented incrementally as redevelopment and enhancement occur.[27]

The Phoenix experience with highway planning suggests that even in communities in which people are firmly committed to their cars, taxpayers may not be willing to pay for increasing highway and freeway capacity. It may be difficult to gain support for good long-range transportation planning if no transportation crisis or terrible congestion is present.

In fact, developing at low densities may keep congestion manageable, reducing the need to provide highway facilities. At the same time, it is clear that the freeway development that has occurred since 1985 in Phoenix has helped limit congestion. If new freeway segments are not built, congestion may again rise to the point where it triggers public willingness to pay for additional freeways.

Notes

1. Tim Schreiner, "Can Phoenix Cope?" *American Demographics*, September 1987, p. 52.

2. Mark Fink, "Toward a Sunbelt Urban Design Manifesto," *Journal of the American Planning Association*, summer 1993, p. 322.

3. "A City Whose Biggest Business Is Itself: Phoenix," *Economist*, May 13, 1989, p. 37.

4. Chris Fiscus, "The Lessons Offered by a Declining Boomtown," *Phoenix Gazette*, May 6, 1994.

5. Michael A. Rosetti and Barbara S. Eversole. *Journey-to-Work Trends in the United States and Its Major Metropolitan Areas, 1960–1990* (Washington, D.C.: U.S. Department of Transportation, 1993) p. 5-3.

6. D.J. Burrough, "Cal Firms Catch Valley Fever," *Arizona Business Journal*, August 4, 1995, p. 58.

7. D.J. Burrough, "Sumitomo '50-50' on Phoenix," *Arizona Business Journal*, July 28, 1995.

8. Kim Shetter, "Sun City Holds On," *Planning*, January 1996, p. 17.

9. Joel Garreau, *Edge City: Life on the New Frontier* (New York: Doubleday, 1991), p. 192.

10. Maricopa Association of Governments, *Urban Form Study*, final report (Phoenix: Maricopa Association of Governments, January 1995), p. 2.

11. City of Phoenix, *General Plan for Phoenix: 1985–2000*, revised version (Phoenix: City of Phoenix, June 1994), p. 1.

12. City of Phoenix Planning Department, *The Phoenix Urban Village Model: General Plan for Phoenix, 1985–2000* (Phoenix: City of Phoenix Planning Department, September 1994), p. 2.

13. Bruce T. Hilby, "Design of Cities No Longer Works," *Arizona Republic*, October 23, 1994, p. E-1.

14. "Phoenix Residents Shoulder Much of Suburbs' Burden," *Arizona Republic*, October 20, 1992, p. A-2.

15. Tom Spratt, "City's Failure to Fight Urban Sprawl Prompts Planner to Quit," *Phoenix Gazette*, August 4, 1992, p. A-2.

16. Lisa Davis, "City Planning Officials Encounter Opposition: Paradise Valley Group Feels Frustrated," *Arizona Republic*, November 9, 1990, p. 5-N.

17. Tom Spratt, "Fighting Decay: Plan Seeks to Save Urban Village," *Phoenix Gazette*, July 16, 1990, p. A-1.

18. Doug MacEachern, "Planning Decisions Give Sense of Directions," *Arizona Tribune*, December 5, 1995, p. 8.

19. Parson Brinckerhoff, *Congestion Management Systems Alternatives*, executive summary (Phoenix: Maricopa Association of Governments, April 1994).

20. *Urban Form Study* (see note 10), p. 5.

21. Maricopa Association of Governments, *Annual Report on the MAG Freeway and Expressway Program* (Phoenix: Maricopa Association of Governments, April 1992), p. 17.

22. Maricopa Association of Governments, *Regional Transportation Planning Report* (Phoenix: Maricopa Association of Governments, September 1989), p. 15.

23. Maricopa Association of Governments. *MAG Freeways/Expressway Plan Update; Cost and Funding Estimates*, executive summary (Phoenix: Maricopa Association of Governments, July 1990), p. 3.

24. Ibid., p. 11.

25. Maricopa Association of Governments, *MAG Freeway and Expressway Program* (see note 21), p. 1.

26. Kathleen Ingley, "Lack of Trust Put Freeway Tax in Pothole," *Arizona Republic*, November 10, 1994, p. A-1.

27. *The Phoenix Urban Village Model* (see note 12), p. 2.

Chapter 7

Transportation and Development in St. Louis
Sprawl without Growth

What we call a traffic jam makes someone from either coast smile with delight. St. Louisans like to say we can get anywhere in 20 minutes.

— Elaine Viets, *St. Louis: Home of the River*

"Picture yourself on a highway over which it is possible to drive at good speed without encountering delays or congestion so common today," a 1947 transportation study for St. Louis invited its readers, "or visualize the possibility of being able to reside in a lovely suburban locality and travel daily to your place of employment in the business district of some large city." For many residents of the St. Louis region, that seemingly naive dream has become a reality. With one of the most affordable housing markets in the nation, excellent residential areas that are located a short distance from downtown, a new light-rail line, and a revitalized economy that is on a steadily upward course, St. Louis would appear to be a true urban success story.

Of course, the reality is a little more complex. Large parts of the city of St. Louis and its older suburbs are in decline. A tradition of small, independent governments focuses leadership on local concerns and makes regional cooperation problematical. Poor people have been left behind in the city—and they lack (affordable) transportation access to jobs in the developing suburban economy.

St. Louis offers a sharp contrast to the other regions serving as case studies in this book. Whereas the other regions have experienced strong growth in population, population in the St. Louis region has been stagnant over the last 25 years (and it actually declined between 1970 and 1985). Population in the city of St. Louis is less than half of what it was in 1950. The city's business base is even more depleted.

Growth management as it is practiced in other cities is a foreign concept in St. Louis, where a spirit of independence and a general distrust of large government works against any organized effort to manage growth on the fringe. In St. Louis, the preferred approach for areas that have experienced growth has been to incorporate as a municipality and then shut down new development.

The St. Louis region has more in common with Cleveland, Pittsburgh, and Buffalo than with San Diego, Atlanta, and Houston. Despite the long-term stability of the region's overall population, portions of the region have experienced a surprising amount of growth—a pattern that has been observed elsewhere in the heartland. The movement of baby boomers into the work force and the growing participation of women in the labor force resulted in a 27 percent increase in regional employment between 1970 and 1990, despite virtually no growth in population. As in other urban areas, increases in driving and a decline in the use of commuting alternatives—especially carpooling—have produced surprising increases in VMT for commuting and other purposes. Population losses in the central area and population gains in the booming suburbs have created demand/supply imbalances in transportation systems. Recent sharp losses in transit ridership appear to have been stemmed, for a while at least, by the introduction of a light-rail system and the restructuring of bus service.

The transportation challenge in the St. Louis region combines the need to provide infrastructure to support the growing portions of the region with the need to maintain deteriorating facilities in the core.

Before the 1970s: The Origins of Regionalism/Antiregionalism

The Seeds of Sprawl

The vitality of St. Louis has always been linked to its transportation system. The city was founded in

1764 near the confluence of the nation's two greatest rivers—the Mississippi and the Missouri. This early competitive advantage has continued through successive transportation developments. The Eads Bridge, the first bridge to span the Mississippi, was built in the late 19th century and provided a connection to the nation's railway system. (It is now used for the region's MetroLink light-rail line.) Charles Lindbergh's operations here brought St. Louis into the aviation age early and started an aviation industry in the region.

In the early 18th century, St. Louis became the seat of government for the northern Louisiana Territory. The rivers were and continue to be an important factor in regional development. Bounded on the east by the Mississippi, the city developed westward in accordance with a plan for streets and roads that was laid out by French and Spanish colonial officials.

The seeds of sprawl were sown early in St. Louis, along with a hands-off government tradition toward managing growth. Since the city's colonial beginnings, St. Louisans have favored a mix of public and private solutions to community problems, and have preferred to keep government close to home, even at the cost of efficient regional government.[1] Absent a clearly enforced municipal agenda, it was the private expansion of transportation infrastructure and the development of new towns that spurred growth in the city. "Real estate developers were fed up with the excessive regulations and insufficient public services inside city limits, so they set up their own towns [along] the waterfront, while settlers spread themselves out across the open countryside west of the border. . . . It was county residents, not city residents, who first asked to separate the city from the county. They were sick of going all the way downtown to the courthouse."[2] As the nation approached its centennial in 1876, St. Louis residents voted to separate the city from the county. This set up a kind of psychological and fiscal wall between city and suburban governments that most other urban regions would not establish until well into the 20th century.

St. Louis developed a streetcar system that "for a time included more track mileage than any in the world."[3] The movie *Easter Parade* immortalized St. Louis's streetcars at the time of the Louisiana Purchase Exposition in 1904. The streetcars served a wide variety of communities throughout Missouri and across the Mississippi into East St. Louis. One of the earliest suburban developments was the city of Wellston, which was served by horse-drawn omnibuses of the Hodiamont line, opened in 1859. By 1876, a vibrant local commercial center had established itself around the loop where the cars turned around.

Motorbus service began in 1923, marking the beginning of the end for the streetcars. The last streetcar line—the Hodiamont line to Wellston—closed in 1966. While many mourned the loss of the streetcars, contemporary news accounts suggest that many transit users preferred the quiet cleanliness of buses to the noise and clamor of streetcars. In addition, bus service could be extended to places where electrically powered vehicles had not been able to go.[4]

Postwar Development

After World War II, St. Louisans, like most Americans, embraced suburbanization with a vengeance. Rationing ended, production workers returned to the assembly lines, and for many the American Dream of a house in the suburbs was realized with the help of cheap mortgages and the construction of the interstate highway system. St. Louis's first interstate highway, I-70, was constructed through the north side, linking downtown with northern portions of the city and county, the airport, and St. Charles County across the Missouri. Eventually, interstate highways grew to 214 miles, making St. Louis's interstate system the sixth most extensive among U.S. urbanized areas in 1990, just behind Atlanta's.

While the expanding highway system was serving a suburbanizing population, some strains were beginning to show in the inner city. The city of St. Louis lost 100,000 people in the 1950s; 128,000 in the 1960s; and 170,000 in the 1970s.

Early Regional Approaches to Development and Transportation

Two initiatives in the 1960s strengthened regional transportation programs. At the national level, the idea that transportation, land use, and economic development are regional—not local—issues was gaining ground, and funding programs were incorporating requirements for intergovernmental cooperation. The Department of Housing and Urban Development established the 701 program to assist local governments in planning for regional needs in housing, water, and other programs that transcended local boundaries. The federal Highway Act of 1962 mandated transportation planning at the regional level. In response to such requirements, St. Louis area governments established the East-West Gateway Coordinating Council in 1965. While the creation of this metropolitan planning organization was an important step in understanding the regional effects of land use and transportation decisions, it did not necessarily change the way business was done. It could not force competing local governments and state transportation agencies to adopt a strategic regional vision. One of the early products of the East-West Gateway

Figure 7-1

Transportation Indicators for St. Louis, 1970–1990

	1970	1980	1990	Percent Change	
				1970–1980	1980–1990
Population	2,410,163	2,376,998	2,444,099	–1%	3%
Commuters	899,598	1,004,504	1,144,336	12	14
In Private Vehicles	749,506	890,557	1,050,392	19	18
Driving Alone	624,806	768,506	975,258	23	27
Passengers	124,700	123,778	75,734	0	–39
By Transit	65,995	55,998	32,318	–15	–42
Commute Share					
Transit	7%	6%	3%	–24	–49
Private Vehicles	83%	89%	92%	6	4
Travel					
Daily Vehicle-Miles Traveled (Thousands)	n/a	n/a	38,618	n/a	n/a
Annual Transit Trips (Thousands)	n/a	75,609	44,342	n/a	–41
Median Single-Family Lot Value	n/a	$15,000	$25,000	n/a	67

Sources: Decennial Census; *Journey-to-Work Trends in the United States and Its Major Metropolitan Areas, 1960–1990* (Federal Highway Adminstration, PL-94-012); and ULI residential land price surveys.

Coordinating Council was a 1969 transit plan, which called for a 100-mile rapid-rail transit system.

The second major regional transportation initiative in the 1960s was the Bi-State Development Agency's assumption and unification of area bus transit services. Transit ridership had been falling and the handwriting was on the wall. John Baine, president of one of St. Louis's last major private transit operators, warned in a 1953 speech that "Americans everywhere should view realistically the time and money they are wasting by driving cars in congested areas, . . . but Americans prefer the luxury of driving their own automobiles." The fragmented transit system was having a complete financial breakdown. Area leaders requested the Bi-State Development Agency—a body that had been established by an interstate compact between Missouri and Illinois in 1949 as a regional development agency with broad powers as a consensus developer—to unify and modernize transit. In 1963, Bi-State purchased 15 operators with a $26.5 million bond issue.

When the Bi-State Development Agency was established, it was given regional planning powers, many of which have since been passed to the East-West Gateway Coordinating Council, and the authority to establish regional transportation operations—including ports, airports, and parking lots—and to develop recreation, refuse, and water supply projects.

The agency has used only a fraction of its broad authorities, but it represents a powerful endorsement of the need for such a regional agency. Being saddled with a regional bus operation did not help its image. Running buses is not a high-profile government activity.

Current Conditions: Economic Diversity and Pockets of Growth

The outsider view that St. Louis is one of the deteriorating Rustbelt casualties of industrial makeover fails to meet the test of reality. The region experienced a solid recovery from the economic recession of the early 1980s, a recession that was deep and long in the Rustbelt. In the process, the economy lost much of its older, obsolete, and less competitive manufacturing, and emerged more diversified and better prepared to face future downturns.

Compared with neighboring states, St. Louis escaped the excesses of oil-based growth and the problems affecting economies with a large agricultural component. During the peak of the good times in the 1980s, the St. Louis real estate market was not as overbuilt as other markets. The worst blows of the early 1990s came not from recession but from cutbacks in defense spending. McDonnell Douglas, headquartered in St. Louis, accounted for one-third of regional job layoffs. Since 1991, regional unemployment has been below national levels.[5]

The St. Louis economy has been on the mend in recent years, with many positive signs. Employment growth in 1994 was the best since 1986, while unemployment hit its lowest rate in 20 years. Retail trade grew at the fastest rate since 1985. New recent revitalization projects include the Kiel Center (a multipurpose auditorium), a convention center, a domed football stadium, and MetroLink. Community self-esteem received a further boost in 1995, when the former L.A. Rams NFL team moved to St. Louis.

St. Louis Compared with Other Metropolitan Regions

The St. Louis region shares many of the characteristics of large urban areas in the Midwest: a stable population, a deteriorating central city, and a heavy dependence on automobiles. The population density of the urbanized area in 1990 ranked 17th among 33 large urban areas, a consequence of the central city's having lost half of its population to low-density housing in the suburbs. In 1990, St. Louis ranked sixth in per capita driving (just behind Milwaukee) and 27th in transit ridership per capita.

The 1990 Census ranked St. Louis third highest among U.S. regions in terms of driving alone for the work trip. Heartland communities occupy all the top spots in this ranking: Detroit and Kansas City precede St. Louis, and Cleveland and Cincinnati follow closely. In keeping with this high level of automobile use, the St. Louis region's extensive freeway system in 1990 had 269 miles, more than Boston's or San Diego's. In number of miles of freeway lanes per capita, St. Louis ranked fourth, behind Kansas City, Atlanta, and Dallas. Its 1,208 miles of arterial streets put St. Louis in 14th place among urbanized areas in terms of arterial road miles—an indication that the region's freeway emphasis is crowding out other road construction.

The region's extensive road system apparently is more than adequate to deal with its traffic growth. Roadway congestion increased by 18 percent between 1982 and 1991, which is about average among large regions studied by the Texas Transportation Institute.[6] Among 15 metropolitan areas with populations between 1.5 million and 3.5 million, St. Louis ranked seventh most congested in 1990, and fifth in the average duration of commute times. Statistically, congestion does not appear to be a problem, and it is certainly not an issue compared with traffic conditions in competitive markets. As Les Sterman, director of the East-West Gateway Coordinating Council, has tried to argue—to no avail, it seems—between 1980 and 1990 the duration of the journey to work increased only one-half minute. However, 51 percent of residents surveyed in 1988 felt that highway congestion was a

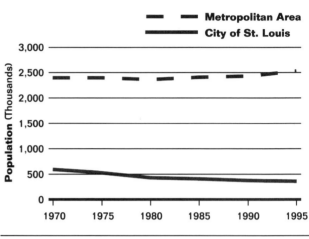

Figure 7-2

Population of St. Louis, 1970–1995

Sources: Decennial Census; and East-West Gateway Coordinating Council.

problem and two-thirds felt that it was getting worse.[7] And some major congestion does, in fact, occur in fast-growing St. Charles County, Missouri, as well as on bridges crossing the Mississippi and Missouri rivers. People's concern about congestion may be related to the region's river barriers. The 1993 flood was a reminder of how much the region depends on good connections.

The region's heavy highway orientation makes it a kind of Houston-on-the-Mississippi. Development of the highway system has kept pace with traffic growth, most of which is due to growth in per capita VMT. The 1990s expansion of the transit system with the opening of MetroLink represents the region's first serious attempt to counteract the loss of the transit market.

Development Market Trends

Population. The city of St. Louis's population losses slowed somewhat during the 1980s. The city has a current population of 366,000, which is only 43 percent of the 1950 peak of 850,000. However, population decline has spilled over to close-in suburban cities. Between 1970 and 1990, 41 out of 91 inner-suburban cities registered population declines.[8] St. Louis County, which had registered a robust gain of 33 percent during the 1960s, stalled at a mere 2 percent increase in population in the 1970s. The region as a whole lost population in the 1970s. The trend reversed during the 1980s, when a slight gain in population returned the region to the population levels of the 1970s.

Employment and the Location of Jobs. While population numbers spelled stagnation, the regional economy was growing and changing. As in many industrial cities in the Midwest, manufacturing has declined in importance, while the role of services has grown. Between 1950 and 1990, about 10,000 manufacturing jobs were lost—an obvious concern, but actually only a small 4 percent loss from the 1950 base of 230,000 manufacturing jobs. Moreover, while a modest loss of industrial jobs has taken place, increased productivity and improved technology have resulted in substantial gains in industrial output, as has occurred throughout the United States.

From 1950 to 1990, the region's employment base grew by 478,000 jobs, leaving the manufacturing sector with only a 19 percent share of the total job market.[9] How is such a large gain in jobs possible with no net gain in population? The movement of baby boomers into the job market and the sharp rise in female participation in the work force are responsible. Between 1980 and 1992, the number of jobs added—140,000—was almost twice the population growth of 75,000. St. Louis felt little impact from the recession of 1990 to 1993 that hit the East Coast and West Coast economies.

All in all, the fact that St. Louis is a growth economy is surprising. Had the region's population increased with economic growth, St. Louis could have faced many of the same problems as other growing regions.

Employment growth is occurring generally outside the urban core. While downtown St. Louis accounted for almost half of the office space in the St. Louis market in 1994, high vacancy rates in the older inventory of office space have kept additions to the 16 million square feet of multitenant office space relatively low. In 1994, office space absorption in Clayton (5 million square feet), which is the county seat, and also in two suburban highway corridors exceeded office space absorption in downtown.[10]

The bulk of job growth has been in business parks and industrial locations in the county. Most of the high-tech research and manufacturing ventures, for example, are clustered along the I-64 St. Louis Technology Corridor extending 36 miles west and 15 miles east of downtown.

In 1990, downtown St. Louis accounted for about 10 percent of regional jobs, and the city as a whole contained one out of every four jobs. Job forecasts, however, indicate a minuscule 1 percent increase from 1990 to 2005 in the city of St. Louis's employment, and only a slightly higher increase for the downtown. St. Louis County is projected to pick up 40 per-

cent of the region's job growth, compared with 23 percent for St. Charles County—just about the reverse of their projected shares of residential growth. The Missouri side of the region is projected to capture about 82 of the region's employment growth, which is just a hair below its share in 1990.

The East-West Gateway Coordinating Council has identified 15 centers that had at least 10,000 jobs in 1990. Downtown St. Louis was by far the dominant center, with 92,000 jobs, but the growing importance of suburban centers is evident. Clayton, the major uptown office center to the west, had 24,000 jobs, while a cluster of activity around Lambert Airport, including Hazelwood and Florissant, had a combined total of 67,000 jobs (see Figure 7-3).

Urban Revitalization. The flood of migrants out of the city in search of newer homes and better schools was initially overwhelmingly white, and it left the poor black population behind. By the early 1970s, the income gap between blacks and whites in St. Louis was among the widest in the nation's cities. As property values declined, the racial composition of inner-city neighborhoods changed. Eventually, there were not enough families to occupy the homes and apartments left behind by out-migrants, so properties became vacant and abandoned. The black mid-

Figure 7-3

Travel to and Jobs in Major Activity Centers in St. Louis, 1990

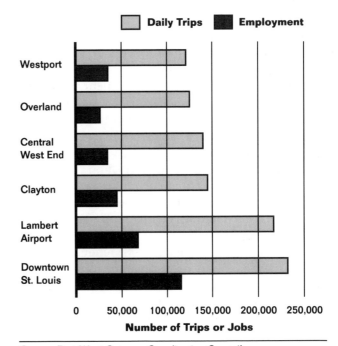

Source: East-West Gateway Coordinating Council.

dle class began migrating to the suburbs in the 1970s, a phenomenon that is still occurring.

The first public response to population loss and impoverishment was traditional urban renewal. A notorious example is the Pruitt-Igoe public housing project, which was eventually torn down. However, the needs were so great that traditional urban renewal did not go far enough, and did not deal with the social problems beyond real estate.

In 1949, the state of Missouri created a new and unique institution: urban redevelopment corporations. This authorized the city of St. Louis to exercise the public's eminent domain authority and abate real estate taxes for projects developed by private, for-profit urban redevelopment corporations. The Missouri mind-set seems to be more comfortable with the idea of creating public programs to stimulate private development than it is with the idea of vesting the government with too much authority.

In 1978, the city of St. Louis created a housing improvement program that is the nationally recognized centerpiece of its public/private partnership approach to urban revitalization. The HIP program offers gap financing loans to low-income rental housing developers. By the mid-1980s, this program combined with other low-income housing programs had helped build or rehabilitate more than 17,500 housing units, which represents an extraordinary accomplishment—although it still falls short of the needs.[11]

The construction of the Gateway Arch in 1965 helped give St. Louis a positive national image at a time when most of the news being reported was pretty gloomy. The construction in 1967 of a stadium in downtown by the Civic Center Redevelopment Corporation, a coalition of major downtown business and financial interests, was an early example of St. Louis style public/private redevelopment, and its success paved the way for over $3 billion in downtown investment, half by redevelopment corporations. (The stadium was subsequently acquired by Anheuser-Busch Companies, Inc.) Examples of other major projects include:

- **Lacledes Landing,** a long-term plan to create an attractive blend of shops, restaurants, and entertainment at the original site of the city of St. Louis, using historic buildings that have been abandoned since the 1960s;
- **St. Louis Union Station,** the 1985 redevelopment by the Rouse Company of a historic train station that had been abandoned in 1976, as a specialty retail center and an adjacent hotel; and
- **St. Louis Center,** the renovation of two downtown department stores and construction of a new retail mall and office buildings to create, in 1985, the

then largest enclosed downtown shopping mall in the nation.

As a result of such redevelopment, employment in downtown actually increased between 1980 and 1990—for the first time since the 1920s—going from about 80,000 in 1980 to over 100,000 in 1990.

In a 1987 evaluation of redevelopment in St. Louis, consultant Richard Ward pointed out that "to attract its fair share of regional investment, a declining central city must be willing to risk its own resources." Ward now notes that "a parochial view drove many of the successes, but that the city will need suburban partners to continue this success. It is no longer possible to act in isolation." Indeed, the suburbanization of decline means that many incorporated and unincorporated suburbs share in the problems that previously were confined to the city of St. Louis.[12] Furthermore, revitalization in the city seems to have slowed in recent years.

Housing. Residential development is strong, and virtually all of it is detached homes. The favored location for new housing continues to be the fringes of St. Louis County and beyond. This locational trend is helped along by bottlenecks in the development process in St. Louis County and municipal incorporations and annexations that are undertaken to escape county planning and zoning regulations and slow down growth.[13] Projections call for the Missouri portion of the region to capture 83 percent of the growth in residential units, slightly above its 1990 share of 78 percent. The lion's share of residential growth is projected for St. Charles County, beyond the Missouri river, which is expected to capture 40 percent of the residential growth between 1990 and 2005, although it contained a mere 8 percent share of households in 1990. St. Louis County, which had more than 40 percent of the region's households in 1990, is expected to experience below-average growth and capture 25 percent of new households by 2015. Forecasts for the city of St. Louis indicate a smaller decline than in the past in population, from 400,000 in 1990 to 348,000 by 2000, with virtually no change in the number of housing units.[14] (Some unofficial projections for the city anticipate much higher population losses.)

Transportation

Employment and housing location trends have affected the transportation system. While the downtown is well served by bus and light-rail transit, Clayton is along a future route and most of the jobs near the airport are more accessible to the nearby interstate than to the airport rail station. The region's new employment centers generate much more traffic—for work as

Figure 7-4

Population Losses and Gains in St. Louis Region by Census Tract, 1980–1990

Census Tracts Losing Population

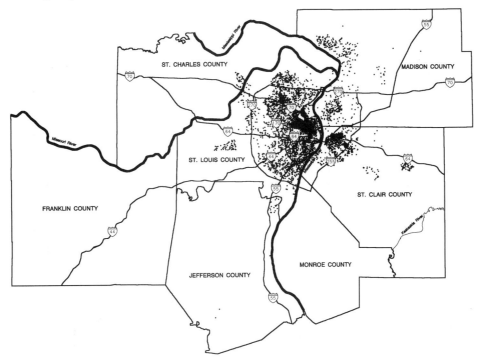

Census Tracts Gaining Population

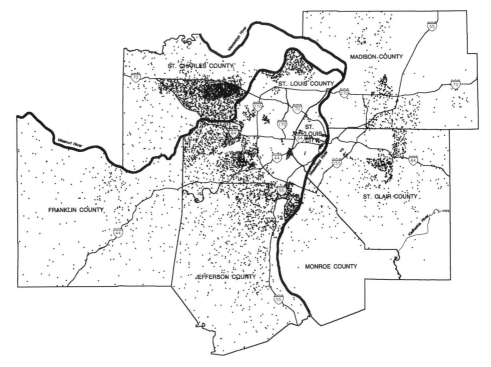

One dot = 30 persons. Dots are randomly plotted by Census tract, but Census tract boundaries are not shown.

Sources: 1980 Census; and 1990 Census. Graphic by East-West Gateway Coordinating Council.

well as shopping and other household travel—than the number of jobs they house, as shown in Figure 7-3.

The Lambert activity center includes the airport, the main campus of McDonnell Douglas, and a Ford assembly plant. It rivals the downtown as a traffic generator. Two other centers—Clayton, the primary suburban office center, and the hospital complex and surrounding intown retail area known as the Central West End—generate more than half as many daily trips as does downtown. Not far behind in traffic generating power are the office/warehouse complex in Westport and the Overland center.

The decline of the downtown as a retail and entertainment destination is evident when these centers are ranked for nonwork trips. In most of these activity centers, the number of nonwork trips exceeds the number of work trips. But this is not true for the downtown, which attracted 15 percent fewer nonwork trips than work trips in 1990. The Lambert activity center generated more nonwork trips than did downtown, and Clayton, which is maturing as a full-service activity center, had three-fourths the number of nonwork trips as downtown.

Regional transportation needs have continued to grow, in part because of the growing economy and in part because of higher levels of motorization. Between 1970 and 1990, the number of workers increased by 27 percent. The number of workers driving to work alone increased by 52 percent. During the same time

period, the number of commuters riding the bus declined by 52 percent, another consequence of the loss of population in the central city and the suburbanization of jobs.

Commuting represents only part of the travel picture. In fact, it is a diminishing portion. Between 1965 and 1990, there was a 50 percent increase in the number of trips made by residents, and the commute between home and work represented only 20 percent of the increased number of trips.[15]

In response to growth in highway travel, the interstate freeway system in both the Missouri and Illinois portions of the region continues to be expanded. However, there has been growing resentment about freeway impacts on established neighborhoods. The construction of I-44 through the city in the early 1970s raised much controversy. A number of neighborhood associations that are now thriving were founded as opposition to highway projects.

Growing antifreeway sentiment led to the city's rejection of state and federal support for a ring road around downtown. Thus, while four interstate highways converge at the downtown, the central business district cannot reap the full economic development benefits of this situation.

The region's network of freeways continued to expand while the inner city declined and the suburbs grew. When it was completed in Missouri in the early 1970s, I-270, which forms a loop 25 miles be-

Figure 7-5

Transit Use by Commuters to Downtown St. Louis by Distance Traveled, 1990[1]

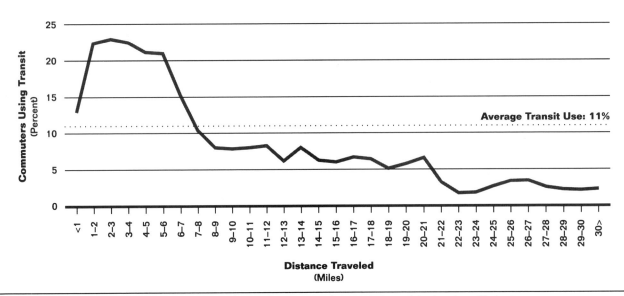

Distance Traveled
(Miles)

[1]The universe is workers 16 years and over who work away from home and commute to work in downtown St. Louis.
Sources: East-West Gateway Coordinating Council; and 1990 Census.

Figure 7-6

Population of St. Louis by Distance from Downtown, 1980 and 1990

1980 1990

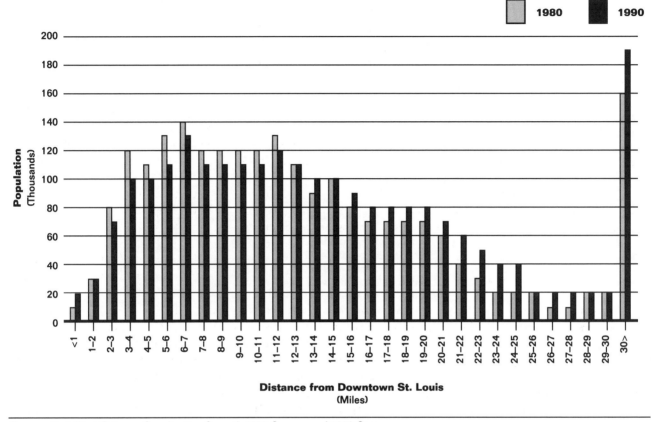

Distance from Downtown St. Louis
(Miles)

Sources: East-West Gateway Coordinating Council; 1980 Census; and 1990 Census.

yond downtown, was thought to describe the outer limit of development. Instead, it has facilitated growth along and beyond it. As shown in Figure 7-4, virtually all of the areas losing population in the 1980s were inside I-270, and all of those areas gaining population were outside it.

Trips into and within the central city represent the prime market for transit. Therefore, the huge loss in population experienced by St. Louis and its inner suburbs has had a devastating impact on transit ridership. In 1990, one out of every five workers in downtown who lived within six miles of downtown rode transit to work, and many others walked (see Figure 7-5). However, the population that lived within six miles of downtown St. Louis declined by 67,000 between 1980 and 1990 (see Figure 7-6). Only in areas 12 to 13 miles out and farther were there gains in net population. In these areas, the transit share of downtown trips was only 6 percent, and residents were much less likely to work in the downtown.

Population losses also have affected transit corridors: along the full length of the initial MetroLink corridor from East St. Louis to Lambert International

Airport, the resident population within one mile of the rail line declined from 159,000 in 1980 to 138,526 in 1990, a loss of one in every eight residents. This did not portend well for the light-rail line that was about to open.

The decline of downtown retail and other service businesses also has taken a toll on the regional transit system. Mirroring the 47 percent decline in central city population from 1960 to 1990, annual transit rides in the region sank from 87 million in 1963, the year that Bi-State took over the region's transit operators, to 37 million in 1992—a 57 percent decline. The number of daily commuters using the bus to get to work declined from 199,000 in 1960 to 32,000 in 1990, a loss of more than five out of every six riders.

St. Louis had the second highest loss of transit commuters among large urban areas in the 1970s, and the highest in the 1980s. Transit's share of total work trips fell from 15 percent in 1960 to less than 3 percent in 1990.[16] St. Louis was by no means unique in the decline of transit ridership.

It is the severity of the physical and economic deterioration in St. Louis combined with larger trends

The Selling of MetroLink

The deteriorating transit market in St. Louis was a problem. During the first half of the 1980s, the number of annual transit riders dropped by more than one-third, from 76 million to 49 million. Fare increases intended to stabilize transit finances were part of the reason. The point of diminishing returns was reached in 1992, when a fare increase caused such a decline in ridership that it failed to produce more revenue. The Bi-State Development Agency realized that a new strategy was needed.

Ambitious plans for a new rail system were in the air. In 1980, U.S. Representative Robert Young organized a publicity trip from the airport to downtown on the abandoned Wabash Railway tracks as a way of pointing out the availability of underused rail rights-of-way and the deteriorating 1884 Eads Bridge across the Mississippi. In 1982, the McArthur Bridge, which crossed the Mississippi between St. Louis and East St. Louis, was closed to vehicle traffic. St. Louis's Mayor Shoemehl created an "underground" bridge group that met in the basement of city hall to develop alternatives. Out of that effort came a plan for establishing rail service on mostly existing rights-of-way. One of the members of the group was the director of the East-West Gateway Coordinating Council, which was given responsibility for carrying the project forward.

A key element of the plan was to count the value of the rights-of-way, including tunnel and bridge, as the 25 percent local match required for federal funding. In 1983, the city traded the city-owned McArthur Bridge for the historic Eads Bridge and downtown tunnel, which were owned by the rail-road. The railroad got a better river crossing, and the city received a bridge and a mile-long tunnel through St. Louis's prime retail, commercial, and entertainment district—a ready-made subway that would have cost hundreds of millions of dollars to construct.

Unfortunately, this deal was not received well in Washington, D.C. The Reagan administration was not supportive of new rail projects. The Urban Mass Transportation Administration head, Ralph Stanley, preferred a bus alternative and refused to accept the value of the bridge and associated property as a local match. The lack of actual cash exposed a serious problem with this proposal, namely, the lack of strong local support, including the support of the transit agency. In addition, one of the prime alignments to Clayton through an upscale corridor was killed by opposition from residents of University City, a close-in suburb, who felt that the planning process had been too heavy-handed. The East-West Gateway Coordinating Council could not very well serve as an advocate, because it was officially conducting a study of different alternatives. Clearly there was a need for grass-roots support, a gap that was filled by an unusual coalition of citizen and business organizations.[1]

In St. Louis, major business interests are represented by Civic Progress, a group made up of the CEOs of the 30 largest St. Louis–based corporations. It was founded in 1953 to help revitalize St. Louis. Many of its members had made corporate decisions by the 1980s to build downtown, rather than in the suburbs—including the Sverdrup Corporation, AG Edwards, and Ralston Purina. The group operates by consensus, and its support for a project contributes considerable financial and political muscle.

Mayor Shoemehl and Hal Dean, former CEO of Ralston Purina, approached Civic Progress to support the transit project. Dean, also a board member of Burlington Northern, was a zealous advocate of the project who had led the negotiations that had resulted in the 1983 deal to swap the bridges. Civic Progress asked him to conduct a study, and the group subsequently gave its support to the initiative for three reasons:

- Rail transit was not only good for the city, but was also in the financial self-interest of a number of business leaders with significant investments downtown.
- It was cheap. No local cash was involved—a far cry from some other proposals with price tags in the billions that Civic Progress had been asked to support.
- It had enough support to convince skeptics not to publicly oppose the project. In fact, most of the opponents thought it would die a natural death.

In 1985, Civic Progress agreed to fund a grass-roots advocacy group in recognition that the transit proposal could not be just a business idea. Business has the muscle, but the public has the votes. A four-year commitment of $300,000 was made to the Citizens for Modern Transit to manage the effort. In addition, individual members of Civic Progress lent their own influence.[2]

Citizens for Modern Transit faced three major challenges: generate a critical mass of grass-roots support that would demonstrate a mandate for

—the suburbanization of jobs and increasing motorization—that has created this worst case scenario for transit and the transit operator, the Bi-State Development Agency.

Outlook: Strengthening Regional Actions

The St. Louis region lacks a concerted public policy to limit the spread of urbanization. Those who would control growth, however, have an ally in the land. Development to the west, in the direction of the region's upper-income housing, has virtually filled in all of the areas within 25 miles of downtown up to the Ozark uplift, a rugged terrain with limited development possibilities. The most promising areas for development appear to be east of the Mississippi in Illinois, which has generally been considered a less desirable location.[17] However, the barrier the river presents to easy travel has now been breached by MetroLink, giving southern Illinois residents a new travel option. Development in close-in Illinois would be advantageous to downtown St. Louis, because it is much closer than the distant Missouri suburbs.

Infrastructure improvements in the region have generally kept pace with development, and the region

The Selling of MetroLink (continued)

light rail; obtain federal funding; and convince the Bi-State Development Agency to get behind the project. An audiovisual presentation was put together, an active speakers bureau made an endless round of presentations to business and civic groups, and materials were prepared for placement in the media. In 1986, CMT was able to divert a light-rail car on its way to San Jose from the factory, which it displayed at Union Station as the feature of a transport weekend. It attracted over 5,000 visitors as well as a lot of media attention. The support of Civic Progress was vital in opening doors to corporate boardrooms and elected officials. CMT organized field trips to new rail projects in Portland and San Diego. Companies would send upper management people, whose enthusiastic reports back reinforced management's support for a rail line. While there were skeptics among citizens and university faculty, in general there was little organized local opposition.

Despite this, Bi-State remained on the sidelines, creating an awkward situation—a rail project opposed by the transit agency. Bi-State was an agency with problems, and an image problem as well: out-of-control expenses and unreliable service. Although its operating expenses were rising more slowly than inflation, the agency moved aggressively to attack costs, through collective bargaining agreements, productivity improvements, the outsourcing of insurance programs, and the renegotiation of utility rates.[3] Bi-State finally turned around with the appointment of two CMT members to its board and the hiring of an executive director who supported the MetroLink project. The region

now presented a unified front for the budget battle in Washington, D.C.

By 1987, MetroLink was planned as an extensive system with links to the north, south, east, and west. The initial line would go from the airport through downtown and cross the Mississippi with two stops in Illinois. A next phase would run east through St. Clair County to a new airport planned at Scott Air Force Base, a route that would improve access from the Illinois suburbs to St. Louis and also assure the support of the Illinois congressional delegation. Polls showed such high grass-roots support that members of the congressional delegation, including Missouri Senator John Danforth and Representative Young, felt that the time was right to push the project in Washington, D.C. All the essentials for making the case with the federal government seemed to be in place: local support, political support, and a favorable draft environmental impact statement.[4] The only problem was that the UMTA administrator, Ralph Stanley, refused to let the project proceed to preliminary engineering with no real cash in the local match. After a year's standoff, however, Stanley resigned, and his successor was a career federal transit official whose confirmation hearings were chaired by Senator Danforth. Soon after the confirmation, the project was approved, and all parties signed off in 1988.

With MetroLink under construction, CMT's initial mission had been fulfilled. The organization then began working to build support for the extensions and to prepare for the opening of MetroLink.

On July 31, 1993, 67,000 people attended the opening ceremonies. This event had been actively marketed by

CMT, at a cost of $700,000. Daily ridership quickly passed the 17,000 forecast for the initial year and reached 44,000 in the 12th month of operation. In level of ridership, MetroLink ranks second among new light-rail systems. At the same time, bus ridership has stabilized. In fact, bus ridership increased by 3 percent during the second year of MetroLink. Perhaps more important, MetroLink has given a psychological boost to the St. Louis region and to Bi-State. Transit's improved credibility has already opened people's wallets. In the fall of 1993, voters in St. Clair County passed a half-cent sales tax to finance the Illinois extension. In 1994, voters in the city of St. Louis and St. Louis County passed a quarter-cent sales tax for improving transit. The revenues that Bi-State has not obtained from the farebox may now be available from the general public, a striking endorsement of rail transit from the "Show Me" state.

Notes

1. Andrew D. Young, *Citizens for Modern Transit: Lobbying for Light Rail in St. Louis* (St. Louis: Citizens for Modern Transit, 1995).

2. Clarence C. Barksdale, "How I Got Business Behind MetroLink: Money, Muscle, and Magic" (paper presented at the Midwest Light-Rail Conference, St. Louis, April 10, 1995).

3. Raymond S. Jurkowski and Nina Thompson, "True Star of the American Road," *Intermodal Transit*, July/August 1994.

4. An early comparison of the rail proposal with a plan to beef up bus service showed that the bus plan would carry more riders than the rail proposal. Cutting through all of the technobabble, the rail proponents asked the consultants a simple question: Is this a bad project? The response was "no," and supporters were relieved.

---

enjoys one of the nation's best highway systems in terms of capacity.

Despite the virtual collapse of the transit market over the years, the transit system is recovering. Metro-Link opened in 1993, to the delight of riders, the pride of residents, and the relief of public officials who had endured criticism that no one would patronize the system. One of the challenges of such an impressive investment in infrastructure is simply keeping it up. Les Sterman, the director of the East-West Gateway Coordinating Council points out that "the single most important determinant of whether the region will be able to shed its Rustbelt reputation and econ-

omy is infrastructure."[18] St. Louis is faced with a twin challenge—how to restore infrastructure in the core while at the same time providing costly new infrastructure in the fringes. This is especially difficult in Missouri, one of the lowest taxing and spending states in the country. Solidly middle-class Missouri spends as if it were poor.[19] Despite this, the public has been receptive to raising money when a clear case for the need has been made, especially for transportation.

MetroLink Strikes a Blow for Transit

The new light-rail line on rail rights-of-way serving the central area was a controversial project from the beginning, not only because some doubted that Missourians would patronize it but also because many lacked confidence in the capacity of the transit agency to build it. But it opened in August 1993, on time and within budget and to a much bigger crowd than had been anticipated. MetroLink has been proclaimed a success, and communities are clamoring for extensions.[20] By mid-1996, daily ridership levels surpassed 40,000. Average weekday ridership during the 1995 fiscal year was 37,000, making MetroLink the second most widely used light-rail line in the United States. The system not only brings commuters downtown, but also offers free midday service between some of the major downtown destinations through special funding from downtown businesses. During the year ending June 1995, the daily ridership in the midday fare-free zone was 4,400 trips. Weekend travel has also proved to be a strong market.

MetroLink has been integrated into the larger regional transit network, with bus routes redesigned to serve the stations. One of the worries was that the rail service would simply divert riders from the bus system. However, surveys show that three out of four MetroLink trips are new to transit. In addition to commuters, the backbone of most transit systems, the rail line appears to have opened up other transit markets—midday trips between downtown offices, restaurants, and shops; weekend trips to the ballpark and the stadium; and trips to the gambling boat in East St. Louis. A social goal of MetroLink was to attract more white residents and suburban residents to transit. Survey data show that this goal has been met: while buses are used by mostly blacks and mostly city residents, MetroLink is patronized mostly by whites and mostly by suburban residents.

MetroLink has also attracted financial support for transit. While virtually all capital costs were paid with federal funds—with the local match being in the form of contributed rights-of-way—local funding was needed for operating costs. Fares pay only one-quarter of the cost of operations. The good feeling that arose from the success of the MetroLink start-up was enough to persuade citizens of St. Louis and St. Louis County to approve a 0.25 percent sales tax in August 1993, by 65 percent and 61 percent, respectively. The tax funds will help support transit improvements, including an aggressive expansion program.

The Bi-State Development Agency has planned an aggressive 25-year capital program for MetroLink extensions, commuter rail, park-and-ride lots, and buses—totaling $1.5 billion, of which $300 million in local funding would match $1.2 billion in federal funds. Preliminary engineering began in 1995 on the first extension, a 20-mile route to the Mid-America Airport in St. Clair County, Illinois. This line will link downtown St. Louis to the eastern part of the region, where much of the growth potential lies. It also satisfies the terms of a political deal that was made to gain federal approval. (Having the Illinois congressional delegation on MetroLink's side was added political ammunition in the quest for federal funds.) Preliminary estimates made in 1995 projected a total cost of $391 million for an alignment along the former CSX railroad right-of-way, or just under $20 million per mile.

While this line is relatively cheap to build, if initial estimates predicting fewer than 5,000 riders daily by 2010 are accurate, the investment will work out to a staggering investment of $163,000 per daily roundtrip rider.[21] This compares to about $18,000 per daily roundtrip rider on the initial MetroLink segment, which cost almost as much to construct, but had almost ten times the ridership. The high cost has been defended as the price of the political deal and as an investment in growth in Illinois.

One other positive development is that the new federal rules make it easier to apply federal funds for transit operating assistance. Transit supporters in St. Louis have complained that the state of Missouri does not provide transit assistance. The state's position, according to Missouri Department of Transportation district engineer Freeman McCullah, is that it has been prevented from spending highway funds on transit in the past, and that when the spending rules changed, funding the backlog of highway projects was a higher priority. It now appears, however, that some of the transportation money that flows through the state may become available for transit.

One other key aspect of transit is the development trends that support and reinforce transit ridership. Demographic trends have placed a real burden on city coffers as well as transit finances. The prime transit markets, city residents traveling downtown and in the central area, have been drastically eroded, while the developing suburban markets of large homes in low-density settings are very expensive to serve by transit. Further encouragement of revitalization in the city and inner suburbs will help build the transit base effectively, adding new ridership revenues at virtually no cost. While there do not appear to be any public initiatives specifically promoting transit-oriented development, in September 1995 the Bi-State Development Agency sponsored a ULI advisory panel on optimizing the community and economic development benefits of MetroLink.

The panel made specific recommendations for three stations, which can be considered prototypes for many others, and Bi-State is working with local communities to develop specific recommendations for transit-supportive development.

Redefining Transportation

In 1992, Confluence St. Louis, an independent citizens research organization, convened an infrastructure task force to evaluate current conditions and recommend a strategy for the improvement of a wide range of facilities in the region, including transportation, water supply, sewers and wastewater, and waterways.

The task force concluded that there needs to be a regional approach to infrastructure planning, which should be integrated with local decisions on land use and economic development and with environmental goals. The task force suggested that implementing agencies, including Bi-State, the regional sewer district, and the state transportation agencies, follow specific planning principles that promote regional goals. In the transportation area, it recommended that the focus should be on maintenance and repair and that the expansion of transportation facilities be concentrated on existing corridors.[22]

At the same time, the St. Louis Regional Commerce and Growth Association (RCGA), the regional chamber of commerce, turned its attention to the economy and saw a need for a more proactive approach to carry the region into the 21st century. RCGA therefore established an economic development planning arm in 1994, the Greater St. Louis Economic Development Council. RCGA reorganized with two operating divisions—one for traditional chamber of commerce functions and one for economic development—and recruited a new chief executive who was experienced in economic development.

In 1995, RCGA produced a strategic plan for economic development. It sets a target goal of 63,500 additional primary jobs (with above-average salaries) on top of the jobs that were predicted to be created, for a total 100,000 net new jobs from 1994 to 2000. RCGA's plan focuses on job creation in trade, tourism and conventions, and technology—in order to capitalize on the region's centrality and diversity of transportation and distribution modes. In support of this development objective, it calls for a range of investments in freight and passenger facilities, among them new bridges across the Mississippi and Missouri rivers, MetroLink extensions, and airport and multimodal rail improvements. Known as the Campaign for a Greater St. Louis, RCGA's initiative is willing to seek new tax funding.[23]

Growth patterns in St. Louis have created a significant problem for transportation investment. Development densities are very low: the amount of developed land in the region has grown at ten times the population growth rate of 35 percent since 1950.[24] Scarce resources must be allocated between new projects in rapidly growing communities on the developing fringe and aging infrastructure rehabilitation projects in established areas. These competing priorities present an equity issue as well. Les Sherman points out that "the biggest problem is the disparity between rich and poor, location and economics. We need to both support the fringe and reinvest in the central city."

According to Freeman McCullah of the Missouri DOT, St. Louis County's lack of authority to manage access to new subdivisions, except those that front on state roads, is a part of the transportation problem. It has precluded the development of a grid of streets to serve a collection and distribution function. The road pattern that has developed forces people onto the main roads, even for local trips. In other states, a collector street network would normally be planned by the county and put in place as part of each new subdivision. The absence of a good local street system appears to be part of the reason for the extensive freeway system: no good alternative is available.

The East-West Gateway Coordinating Council has picked up on a theme—sustainable development—that has great appeal, and given it a regional spin: meeting the needs of the local community without jeopardizing the quality of life of any other community in the region. Its director, Les Sherman, acknowledges that this is a challenge in the St. Louis region, where there is not much support for growth management. The sustainable development theme has a practical angle: there simply is not enough money to continue to just build more to meet the region's infrastructure needs.

Within the region, there are many inconsistencies between transportation decisions and development decisions. McCullah cites a few examples. A bedroom community is growing up in St. Charles County, but all of the jobs for its residents are across the river in St. Louis. An expensive new bridge is being planned with no consideration being paid to factors such as increased parking charges downtown, tolls, or growth restrictions. A community in St. Louis County incorporated in order to override county zoning that would have allowed houses on quarter-acre lots.

The state of Missouri is trying to deal with transportation/land use conflicts through better management and maintenance of transportation facilities. It supports the promotion of infill development in areas where infrastructure is already available. It has established a program to spot incidents on the roads and quickly clear them to avoid further congestion. It is rethinking maintenance and reconstruction standards and practices. For example, rather than deferring maintenance on a road planned for rebuilding, the state might undertake a number of small improvements—pavement overlays or limited lane additions

—that can make traffic flow better now and possibly reduce rebuilding needs.

A close look at the St. Louis region reveals that it may be a model for communities that are currently thriving but may soon have to face the twin challenges of growth and deterioration: providing transportation facilities to serve growth and reinvesting in the aging infrastructure in older urban areas. The region seems to have coped so far by doing a little bit of everything:

- Radial and circumferential highway facilities in the exurbs, to serve the growth that is taking place because of market pressures as well as antigrowth actions in less distant municipalities;
- Office and industrial development along these same exurban highways, to serve workers already living in the area;
- Residential and commercial revitalization in the city, especially in the downtown, through public incentives for private development efforts; and
- Major projects—bridges and MetroLink—to provide relief at bottlenecks.

As the region works toward the goal of developing 100,000 new jobs, its transportation policies and growth management policies will need a more strategic framework and a more effective regional focus.

One of the hallmarks of policy making in St. Louis has been its project focus. Each project generates a need for a new coalition. The trick in a politically fractionalized region has been to put together enough different interests to enable an idea to overcome the natural political inertia. Within the city of St. Louis, projects typically require a mix of white and black support. Elsewhere, projects require different mixes of support—central city and county, or suburban cities and suburban counties, or state and local. A premium is put on finding projects that have the critical depth of support to pass, rather than on projects that are best for the region.

To reduce the focus on political deal making, the East-West Gateway Coordinating Council has developed a new approach to transportation planning that focuses on regional goals and objectives and involves three investment criteria—safety, congestion management, and sustainable development. Customer satisfaction, measurement, and feedback are also key elements. However, when matters get down to hard decisions on the allocation of scarce financial resources, some backsliding seems to take place. Everyone agrees that the preservation of existing transportation facilities is essential. But, according to current projections, it would cost $12.6 billion to maintain road and bridge facilities between 1995 and 2015. This would leave only $695 million in the budget for all new construction through 2015. The region's recommended plan—which, in accordance with federal rules, is fiscally constrained—seems to accept the deferment of 9 percent of needed maintenance in order to squeeze out enough new projects to keep everyone happy. (Even this compromise, Sterman points out, represents a substantial improvement over current levels of funding for maintenance.) The plan hints that the states of Missouri and Illinois could bring additional resources to bear, since they have the discretion to move money between different highway districts.[25]

The success of the first MetroLink line complicates the transit financing picture. Now everyone wants one, and the plans are accommodating. Bi-State's capital investment plan calls for light rail along five additional corridors—in addition to the St. Clair extension into Illinois that will be financed with local revenues—over the next 25 years, as well as commuter-rail and bus improvements. It is estimated that $300 million in local funds can finance a $1.5 billion program. The East-West Gateway's latest regional transportation plan, *Transportation Redefined*, hedges on transit financing. It assumes no new rail lines beyond the currently planned extension in Illinois, deferring to the results of forthcoming corridor studies. Spending for transit and operating subsidies is expected to account for 27 percent of total spending for transportation, which, considering that transit ridership accounts for less than 2 percent of daily trips, is a large share.

Transit investment also faces the new areas/old areas choice: spending to reinforce existing services versus spending to extend transit to new markets. St. Louis's answer is: do both. However, heavier competition from other cities for federal funds to finance new projects at the 80 percent level is likely to limit funding for new construction.[26] Bi-State has lowered its expectations for new federal funding. According to Susan Stauder, special assistant for strategic planning, the agency expects no more than 50 percent of the capital costs to come from the federal government.

In 1995, the agency initiated major planning studies for three radial corridors and for a cross-county route that could include commuter rail. It has also begun preliminary engineering on the St. Clair extension. Planners are investigating less expensive construction options for the St. Clair extension, as well as aggressively seeking new funding options and revenue sources, including revising federal rules to make private financing more attractive.

Stauder identifies a change in people's attitudes regarding transit funding. Having once considered transit investments as a regional tax that supported have-nots in the city, suburbanites have become more willing to support transit—and other regional institutions like museums and the symphony—in part because they use them. More than half of MetroLink riders come from the suburbs, giving suburbanites a self-interest in transit as well as a greater willingness to help the city.

However, an August 1996 referendum on extending MetroLink into St. Charles County was defeated. The defeat may have been a result of low turnout, or it may indicate some flagging in the region's initial enthusiasm for rail extensions. Another vote is planned for November 1996.

The regional transportation plan's call for more regional participation in decisions is not surprising, considering its source. Unfortunately, the outward spread of development is probably weakening regional ties, which were not that strong to begin with. The more distant communities probably feel essentially uninvolved with St. Louis and the inner suburbs.

On the positive side, late in 1995, the Missouri Highway and Transportation Department decided to become a full partner in the regional process. Because so much of the region's transportation funding comes through the state—projections for 1995 to 2015 indicate $4.3 billion in state funds, $4.8 billion in federal funds directed through the state, and $4.2 billion from local sources—the state represents the 300-pound gorilla. It—and its counterpart in Illinois—can pretty much call the shots.

In the past, the East-West Gateway Coordinating Council (the MPO) and the state viewed each other with suspicion. The state would deal on individual projects with individual local governments, playing to provincial interests rather than regional concerns. Freeman McCullah says his job "used to be to tell St. Louis what Jefferson City (the state capital) said it can do."

Now, when the transportation department is approached by a city or suburban government about a sizable project, its response is to take it to the MPO. This change in attitude was one of the goals of the federal ISTEA legislation. Jim Farrell of RCGA points out that his organization helped as well, by bringing public officials and the business sector together to examine the needs. The state's leadership came away from the needs assessment with an important message: it needed additional funding in order to catch up; it could no longer go it alone; the state would need the support of the St. Louis community to raise money. It may be hoped that strengthening the regional planning organization will make competing local governments confront some of the difficult growth and land use problems that make the transportation problem worse.

Lessons

Some of the lessons from St. Louis have broad applicability to other older urban regions, and even to some newer regions. They include the following:

No growth in population does not mean no transportation needs. St. Louis has one of the worst cases of a declining central city. Regional population growth has been virtually stagnant, while some suburban communities have grown enormously. On the regional level, VMT has grown considerably despite the lack of population growth, and in some communities travel gains have been explosive.

There may be untapped transit markets in older communities. While most new transit projects are oriented toward high-growth communities, MetroLink has found a new market for access to downtown. Piecing together abandoned rail lines does not necessarily serve many people, but in this case the approach was a winner as it managed to connect downtown (jobs, restaurants, shops, and tourist attractions), the airport, and universities. In effect, MetroLink created heavy-rail service at light-rail prices. MetroLink appears to afford opportunities for infill development, which, by also supporting the transit investment, would serve two objectives.

Despite its large and successful transit investment, this older midwestern region is as automobile dominated as faster-growing Sunbelt cities. St. Louisians are more likely to drive to work than are Angelenos. Regional growth will bring more drivers to highways than it will attract new riders to transit.

Lack of strong regional structures does not preclude regional functions. Pervaded by intergovernmental jealousy and competition, the St. Louis region represents an extreme of antiregionalism in the United States. This culture was blessed in the last century and is unlikely to change soon. However, antiregionalism does not prevent effective regional cooperation around specific objectives. A regional waste management district has been operating effectively since 1953. Other regional initiatives include the zoo, the junior college system, and the transit agency, which recently pulled off the construction of MetroLink's first installment. St. Louis proves that you don't have to like one another to cooperate, if the need is important. Ad hoc cooperation may be an imperfect solution, but it works.

Notes

1. Catherine Corbett, "St. Louis History," in *St. Louis Currents: The Community and Its Resources* (St. Louis: The Leadership Center of Greater St. Louis, 1992).

2. Eric Sandweiss, remarks at a conference (East-West Gateway Coordinating Council, St. Louis, October 1, 1992).

3. Ibid.

4. Ben Katz, *One Hundred Years of City Transit in St. Louis,* Occasional Paper, no. 3 (St. Louis: National Museum of Transport, March 15, 1961).

5. Richard C. Ward and Robert Lewis, "Economy and Employment," in *St. Louis Currents* (see note 1), p. 22.

6. *Trends in Urban Roadway Congestion—1982 to 1991* (College Station, Texas: Texas Transportation Institute, 1995).

7. East-West Gateway Coordinating Council, *Perceptions of Transportation Issues in the St. Louis Metropolitan Area: A Survey* (St. Louis: East-West Gateway Coordinating Council, 1988).

8. East-West Gateway Coordinating Council, *Transportation Redefined: A Plan for the Region's Future* (St. Louis: East-West Gateway Coordinating Council, 1995), p. 17.

9. From materials provided by the Bi-State Development Agency, East-West Gateway Coordinating Council, and Washington University's Urban Research Design Center, for a briefing book used by an Urban Land Institute advisory panel on MetroLink, which met September 17–22, 1995, in St. Louis.

10. Nooney Krombach Real Estate Services, *Market Trends: 1995* (St. Louis: Nooney Krombach, 1995), p. 3.

11. A 1972 study estimated that, between 1970 and 1980, a total of 45,000 new housing units would be needed. See *Urban Decay in St. Louis* (St. Louis: Washington University Institute of Urban and Regional Studies, 1972).

12. Richard Ward (president of Development Strategies), interview by author, February 2, 1996.

13. Development Strategies, Inc., "St. Louis Metropolitan Area," in *ULI Market Profiles: 1995* (Washington, D.C.: ULI–the Urban Land Institute, 1995), p. 263.

14. *Transportation Redefined* (see note 8), pp. 202–203.

15. East-West Gateway Coordinating Council, *St. Louis Region Small Sample Travel Survey,* executive summary (St. Louis: East-West Gateway Coordinating Council, March 1991).

16. Transit estimates for different years are reported in Michael A. Rosetti and Barbara S. Ebersole, *Journey-to-Work Trends in the United States and Its Major Metropolitan Areas, 1960–1990* (Washington, D.C.: U.S. Department of Transportation, 1993), tables 5-9, 5-9A.

17. Ward and Lewis (see note 5), p. 21.

18. Les Sterman, "Regional Infrastructure," *St. Louis Currents* (see note 1), p. 101.

19. Donald Phares, "Financing Local Governments," *St. Louis Currents* (see note 1), p. 43.

20. What constitutes the success of a transit project is open to a number of interpretations. Clearly, the region's residents, elected officials, and the media consider the the MetroLink system to be a success. Bi-State had projected 13,000 riders at first, and 17,000 by the end of the first year. Daily ridership in the first year averaged 24,400 and peaked at 44,000 in July 1994, garnering praise even from critics. Daily ridership increased to 37,000 in the second full year, when an extension to the airport was completed. But, the initial estimates of ridership might have been deliberately conservative in order to guarantee favorable initial reviews. A fair comparison of results would be with forecasts used in deciding whether to build the rail system. Checking numbers against longer-term forecasts requires picking through a minefield of technical data. The final environmental impact statement on MetroLink that was submitted to the U.S. Department of Transportation in 1987 projected 37,000 daily trips by 2000, which is a target already exceeded. A draft EIS prepared in 1984 forecast daily ridership of 39,000 and 46,000 respectively for two alternative rail alignments for 1995. That study also predicted that total daily bus and rail ridership would be 172,000, which is lower than could have been achieved with a beefed-up bus system alone. The forecast for combined bus and rail ridership for the second year was 166,000, which is only slightly below the 1984 projection of total system ridership. All things considered, it appears that MetroLink ridership levels at least match those that were anticipated during early planning.

21. Assuming an average of two daily trips per rider, the projected daily ridership of 4,847 works out to 2,423 roundtrips. Ridership and cost estimates are as reported in U.S. Department of Transportation, *Report on Funding Levels and Allocations of Funds* (Washington, D.C.: U.S. Department of Transportation, May 1995), p. A-156.

22. Confluence St. Louis, *A Vision for Regional Infrastructure Improvement* (St. Louis: Confluence St. Louis, September 1992).

23. Greater St. Louis Economic Development Council, *Strategic Plan for Economic Development* (St. Louis: Greater St. Louis Economic Development Council, February 1995).

24. *Transportation Redefined* (see note 8), pp. 89–90.

25. Three projects—the Page Avenue extension in Missouri and a bypass and a highway relocation in Illinois—make up 71 percent and 80 percent of the capacity costs for the respective portions of the region, and "capturing additional funds just for those three projects would virtually eliminate the need to defer any preservation." See *Transportation Redefined*, p. 161.

26. *Transportation Redefined*, p. 164.

Chapter 8

Transportation and Development in Toronto

A Pioneering Transit Model in a Suburbanizing Future

Toronto found a way to use Metro as an instrument of equality. That's why our core is not the hole in the doughnut.

— David Crombie, former Toronto mayor

Toronto offers a number of firsts in planning, development, and transportation that have made it a model for planners throughout North America.

- Toronto decided to build subways at a time (1948) when everyone was building freeways.
- Toronto was the first metropolitan region to establish a regional government with centralized land use and transportation decision-making powers (1953).
- Toronto was the first region to dedicate local taxes for the construction of rapid-rail transit (1964).
- In 1972, Toronto adopted a transportation funding program that favored transit over roads.
- In 1975, Toronto began investing in transit as a strategic tool for influencing land development.[1]

In addition—and, many say, because of these accomplishments—Toronto is one of the most admired among the world's cities. It was ranked fourth best place to live in the world (1994)[2] and fifth most desirable location for an international headquarters (1993).[3] One of the best endorsements comes from Jane Jacobs, the noted urban historian, who chose to relocate to Toronto.

The region has been a model for other communities because of its early adoption of transportation, housing, and governance policies designed to preserve and enrich the central city. Recently, however, market forces have made decentralization an issue in Toronto, challenging the transportation system and the regional governance structure. The political climate has changed as well. Toronto is in search of new mechanisms or old solutions in new forms that will help it to meet new challenges and possibly make it again a leading model for modern cities and regions.

Before the 1970s: Transportation Planning Pioneer

Early Development

In 1749, French fur traders established Fort Rouille in present day York, and a series of conflicts with the British ensued. The area began to grow following the cessation of hostilities in 1814, and was incorporated as Toronto in 1834. Travel through Lake Ontario and the coming of the railroad in the late 19th century helped the city begin to boom. The first plank road in Canada was built east of Toronto in 1835. This marvelous innovation so impressed the commissioners who controlled Yonge Street north of the downtown that they followed suit on portions of their road.[4] Toronto's first omnibus service began in 1849. The privately operated service carried passengers between the St. Lawrence Market and the city of Yorkville in six-person carriages pulled by horses. A streetcar system—Canada's first—began operations in 1861. In 1892, the number of riders amounted to 26 percent of the labor force. That same year, the system was converted to electric traction.

Toronto boasted 522,000 inhabitants by 1921. The tramways shaped much of the new development occurring at this time, although Canadian tramway companies were generally not involved in real estate development like some of their U.S. counterparts. Electric tramways simultaneously dispersed population and concentrated economic activity.[5] In addition to suburban lines, 25 intercity electric railway companies operated in Canada during the peak year of 1916. Most intercity lines provided access to the downtown, but two Toronto companies served primarily suburban areas.

The impact of automobiles—adopted later in Canada than in the United States—began to be felt on transit ridership. In Canada in the 1920s, a period of growth and prosperity, tramway ridership grew by only 3 percent. A downtown subway was proposed for Toronto in 1910, but voters disapproved it. When the Bloor viaduct was built across the Don Valley in 1918, it was designed to support a subway line below the bridge deck. (This advance thinking paid off almost four decades later, when the Bloor-Danforth subway line was constructed.[6])

The city of Toronto established the Toronto Transportation Commission (TTC) in 1920. The following year the TTC took over and integrated nine existing transit systems. Between 1921 and 1953, it added 35 routes and extended 20.

Pressure for highway improvements in Canada antedates the appearance of the automobile. The Ontario Good Roads Association, formed in 1894, represented farmers, county engineers, merchants, and bicyclists in pressing government for better roads. In 1901, the Ontario legislature passed the Highway Improvement Act, making grants available to county governments for roadbuilding. Federal funding was not available for road improvements in Canada. An Ontario Public Works and Highways Department was established in 1917, but it lacked a coherent highway program until 1920, when federal legislation required it to develop one. (By the late 1920s, all provinces had established highway departments and embarked on improvement programs, which generally focused on the paving of unpaved roads.) Between 1919 and 1936, highway spending in Ontario accounted for 43 percent of all highway spending in Canada. The 1939 opening of the Queen Elizabeth Way (QEW) between Toronto and St. Catharines, at 61 miles the longest divided highway in the world, brought Canada into the era of high-speed highways. Provided with lighting for most of its length, the QEW integrated the concepts of parkway and high-speed controlled-access highway, incorporated high levels of design and landscaping, and included the first grade-separated cloverleaf in Canada.[7]

Having subsidized roadbuilding and the Canadian Pacific Railway, federal officials were initially reluctant to subsidize aviation as another transpor-tation system serving Canada. The federal government encouraged municipal support of airports through airmail contracts and a flying club program to train pilots. The Toronto flying club was organized in 1927. It operated out of an air force field until a club airport was built. Malton Airport began as a one-story brick building adjacent to a wood terminal that had been built in 1939. (Its redevelopment in 1946 into Toronto International/Lester B. Pearson International Airport has been described as "the single most ambitious project in the transition to the jet age."[8]

Regional Services and Transportation Planning

Following the Second World War, residents of Toronto joined the rush to buy cars and move to the suburbs. Vehicle ownership increased by 350 percent between 1947 and 1973. The high cost of housing in the urban area supported the flight to the suburbs. Support for public transit waned, and cutbacks in transit service to save money resulted in fewer riders and lower revenues. The density of new suburbs was half the density of the inner city, which made serving them by transit difficult. It was assumed that the private auto would be dominant and the long-range plans of the time perpetuated that assumption. Toronto developed plans for extensive networks of freeways and began to implement them.[9] One significant difference between Canadian and U.S. urban transportation policies of the period was that freeway initiatives in Canada were led primarily by local and provincial officials. The federal government had virtually no role.

In 1953, Toronto took two major steps to integrate government services. These two actions have been credited with much of Toronto's success in the efficient delivery of municipal services while retaining a strong regional vision. The province of Ontario created the Municipality of Metropolitan Toronto, a federated system of the 13 municipalities in the area (a number that has been reduced to six through consolidations), in order to provide more effective delivery of many government services. Metro Toronto was given responsibility for regional services, including strategic land use planning (which includes land subdivision), transit, and regional roads. The municipalities retained responsibility for local services, including fire, water, local libraries and parks, and local land use planning and zoning. Although the creation of Metro Toronto was a major step toward a regional government, the legislation stopped short of giving zoning control to the regional agency. The same legislation created the Toronto Transit Commission, formerly the Toronto Transportation Commission, with monopolistic authority over integrated transit

Figure 8-1

Population of Toronto, 1971–1994

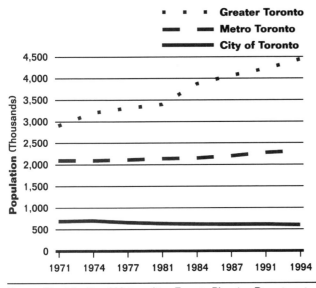

Legend:
- ▪ ▪ ▪ Greater Toronto
- ━ ━ Metro Toronto
- ━━━ City of Toronto

Source: Municipality of Metropolitan Toronto Planning Department.

services—trolley, bus, and subway (the first segment of which was under construction).

Despite its declining influence, transit still played a significant role in Toronto. Streetcar systems had been dismantled in every Canadian city, as well as in most U.S. cities, except for Toronto. In 1948, the city of Toronto decided to build the Yonge Street subway line—as a cost-effective alternative for several heavily used streetcar lines and as a way of freeing up road space for cars. The first section was completed in 1954. It had been financed without debt because the transit agency had accumulated a large cash surplus during the war, when automobiles and gasoline were scarce.

During the 1960s, the Yonge Street line was extended, a Spadina-University line parallel to Yonge was constructed, and a crossline was built on Bloor Street, serving suburbs to the east and west. The region was growing strongly at this time. Population rose from 1.6 million in 1961 to 2.1 million in 1971. The subway supported the growth of the downtown[10] and offered attractive locations for high-density apartments along its route. A metropolitan apartment control policy encouraged apartment construction within 1,500 feet of subway or commuter-rail stations and restricted development in areas without transit access. Between 1965 and 1974, when average annual apartment construction was running at 16,000 to 17,000 units, the pull of subway sites combined with the push of public policy helped create high-

rise clusters around the stations, primarily on the north/south Yonge line.[11]

Regional transportation planning was a developing science in North America in the early 1960s. Some of the leading models of the time included Chicago, Detroit, and Pittsburgh. Metro Toronto's planners pioneered the development of some analytical traffic forecasting tools that have been adopted by planning agencies throughout the world. The Traffic Research Corporation, the Toronto firm that developed Metro's computerized traffic forecasts, had also developed a computer model for predicting land use in eastern Massachusetts.[12] Canadian and U.S. transportation planners shared not only planning tools, but also transportation philosophies. A 1966 plan adopted by the Metro Toronto planning board called for a balanced transportation system of 30.5 miles of subway and 105 miles of expressway. It sparked a freeway revolt. Neighborhood, environmental, and antigrowth organizations opposed the expressways. As each expressway was deleted from the plan, "Metro's planners and politicians fought a rearguard battle to save the remaining elements, even though the network made less and less sense."[13]

Current Conditions and Recent Initiatives: A Transit Orientation in Travel and Planning

Toronto Compared with Other Metropolitan Regions

Ranking Toronto among U.S. urban regions for various density, form, and travel indicators, while interesting, fails to take account of the sharp differences between cities in the two countries. Canadian cities and regions have higher density, greater transit service and use, higher multifamily occupancy, and generally lower levels of auto ownership. It is more appropriate to compare Toronto with other Canadian urban areas. A survey by the Transportation Association of Canada makes such comparisons possible.

Greater Toronto is the largest urban region in Canada. In 1991, it had 4 million people, compared with Montreal (3.5 million) and Vancouver (1.8 million). Among U.S. regions, Greater Toronto is slightly larger than Boston, and only six regions are larger. Its growth rate of 23.9 percent during the 1980s was slightly above Seattle's. Only nine U.S. metropolitan areas grew faster.

The density of Toronto's existing urbanized area is about 6,400 persons per square mile, about equal to that of Montreal. Toronto's density is about 29 percent greater than that of the New York urbanized area and also greater that of than Los Angeles, which is the densest U.S. urbanized area.

In 1991, Toronto ranked second among Canadian regions in per capita transit ridership, second to Montreal. With 135 per capita trips annually, Toronto was about equal to Ottawa, below New York, and 7 percent above Boston, the second most transit-oriented U.S. region.

The Toronto region ranks high on transit use during the morning peak period, when 20 percent of all travelers use transit—compared with 27 percent in Montreal. In walking, Toronto ranks well behind Ottawa and Quebec. Walking's share of the morning commute is a low 4 percent, an issue planners hope to address. With 19 percent of commuters using transit in 1991, as shown in Figure 8-2, Toronto ranks behind only New York among top ranked U.S. regions for transit commuting.

Estimating the amount of driving in Canadian cities and metropolitan regions is difficult, because of the way statistics are maintained. No federal/state statistical program like the U.S. Department of Transportation's exists. The Transportation Association of Canada is trying to fill data gaps with its own surveys. In Toronto, it is estimated that in 1991 the number of daily miles driven on arterials and expressways per resident was 10.1, which was about 14 percent less than similar miles driven in Ottawa. Compared with U.S. urban areas, daily per capita driving in Toronto is about two-thirds that in New York and about half that in Boston, San Francisco, and Washington, D.C.—all regions that rank below Toronto in transit use.

A surprising finding is that more automobile travel in the Toronto region was on expressways than on arterial streets. When ranked by the supply of arterial and expressway lane-miles per thousand inhabitants, Toronto, with about 2.2 miles, ranks behind Vancouver, but well ahead of Montreal, Ottawa, and Quebec.[14] Compared with freeway mileage in U.S. urbanized areas, Toronto's freeway system has relatively few lane-miles. The region joins Chicago and Philadelphia at the bottom end of this ranking.

A comparison between Toronto and Denver (see Figure 8-3) reveals the kind of dramatic differences that make meaningful comparisons of transportation and development indicators in U.S. and Canadian cities virtually impossible. Both regions have populations in the 2 million range. Metropolitan Denver extends over an area that is almost ten times the size of Toronto. Despite incomes that are higher than the national average in Toronto and lower than the national average in Denver, per capita transit use in Toronto is almost six times transit use in Denver. This has financial consequences. In Denver, only 21 percent of operating costs are covered by fares; in Toronto, fares cover 71 percent of such costs. Although comparisons are difficult, U.S. and Canadian regions may share some common lessons.

Figure 8-2

Transportation Indicators for Greater Toronto, 1971–1991

	1971	1981	1986	1991	Percent Change 1971–1981	1986–1991
Population	2,920,250	3,417,701	3,733,060	4,235,756	17%	13%
Employment	1,030,050	1,817,130	2,029,055	2,216,675	76	9
Commuters	n/a	n/a	1,726,151	1,880,864	n/a	9
In Private Vehicles	n/a	n/a	1,255,368	1,433,163	n/a	14
Driving Alone	n/a	n/a	1,100,436	1,279,840	n/a	16
Passengers	n/a	n/a	154,932	153,323	n/a	−1
By Transit	n/a	n/a	384,384	360,141	n/a	−6
Walking	n/a	n/a	71,198	68,111	n/a	−4
Other	n/a	n/a	15,201	20,854	n/a	37
Commute Share						
Transit	n/a	n/a	22%	19%	n/a	−14
Auto	n/a	n/a	73%	74%	n/a	1

Source: The Transportation Tomorrow Survey (Municipality of Metropolitan Toronto Planning Department, 1995).

Figure 8-3

Toronto and Denver Compared, Early 1990s

	Denver Region	Metro Toronto
Area (Square Miles)	2,300	240
Population (Millions)	1.8	2.2
Average Density (Persons per Square Mile)	780	9,200
Average Income	below national average	above national average
Daily Transit Passengers (Thousands)	173	1,500
Annual Transit Trips per Capita	28	162
Share of Transit Operating Costs Covered by Fares	21%	71%

Source: "Rethinking Urban Transportation" (Richard M. Soberman, 1996).

Development Market Trends

During the 1970s, population growth slowed sharply in Metro Toronto. While Ontario grew by 12 percent, the region grew by a mere 2 percent. The slight net increase in regional population was made up of a large gain in the outer areas that counterbalanced a large decline in the inner neighborhoods. The Greater Toronto Area (GTA) includes Metro Toronto and four outer regions: Durham, York, Peel, and Halton. In the 1970s, the four largely undeveloped outer regions grew rapidly, adding almost 500,000 people, which was almost ten times the growth in Metro Toronto.

In the 1980s, population growth in Metro Toronto recovered somewhat, while the surrounding region continued to expand. The outer tier reached 2 million people in 1991, equal to Metro Toronto's 1971 population.[15] It accounted for 83 percent of Greater Toronto's population growth from 1981 to 1991.

Development in these auto-oriented outer suburbs tends to be single-family homes built at much lower densities than are found in Metro Toronto (although still high by U.S. standards). The grafting of American style suburbs onto compact, transit-oriented Metro Toronto (which, however, according to a 1989 survey, more than half its residents are willing to leave because of poor transportation) is called "Vienna surrounded by Phoenix" by transit planner Juri Pill.[16]

The city of Toronto is only one of four cities in the region with a population in the half-million range: Toronto (600,000), North York (542,000), and Scarborough (485,000) in Metro Toronto and Mississauga (433,000) in the neighboring region of Peel. Other cities in the region and outside it are likely to grow faster than Toronto in the future. In 1996, first-time homebuyers accounted for 60 percent of the for sale housing market. Many of them are attracted to suburban areas where prices are below the regional market's high average price of $202,000.[17]

As Canada's economic and financial center, Toronto is home to nearly 40 percent of the top 500 Canadian companies. *Fortune* magazine has named Toronto as one of the top ten cities in the world in which to conduct business. The area also has a multicultural character and attracts a disproportionate share of immigrants to Canada. New immigrants are expected to number 80,000 to 90,000 annually for the rest of the decade. The Toronto metropolitan area holds one-quarter of the country's population and 15 percent of its jobs.

The Canadian economy has historically been tied to the United States, its main trading partner. Metro Toronto was hardest hit among Canadian cities by the last deep recession to hit both countries, and it has been the slowest to recover. The metropolitan area lost 200,000 jobs during the recession.

The Toronto economy is the dominant component of the highly successful provincial economy. Ontario has absorbed a rapidly expanding labor force with low unemployment, and has maintained a per capita output that is consistently 10 percent above the Canadian average. As in most other industrialized economies, the service economy has increased sharply, while manufacturing has declined. In Metro Toronto, this shift has been most pronounced. Not only has manufacturing's share declined, but also the total number of manufacturing jobs has fallen.

A downtown office building boom had already produced an oversupply of space when the recession stalled office jobs in 1989.[18] Office markets were recovering in 1995, although concerns lingered over government layoffs and corporate downsizing. In downtown, as the number of workers declined, increased competition caused parking rates to drop, which put additional pressure on transit.

During the 1970s, the outer regions gained employment as well as population, although half of the new jobs were located in Metro Toronto—a large percentage of them in the downtown Yonge Street corridor. As downtown became more focused on financial, corporate, and government employment, manufacturing and industrial jobs moved to the outer suburbs, which offered large parcels of vacant land with good freeway access and free parking.

Historically, downtown and midtown Toronto have been the dominant office employment areas.

Beginning in the early 1980s, the suburbs began to capture a greater share of the office market.[19] Suburban municipalities offered generous economic incentives to attract businesses. They have capitalized on the property tax rate differential between the downtown core, where office properties are taxed $10 per square foot, and the suburbs, such as Mississauga and Pickering, where commercial real estate taxes are $3.10 per square foot and $1.66 per square foot, respectively.[20]

The coming of the Toronto Raptors, a new NBA franchise, and the expansion of the convention center are expected to help revitalize the downtown. Another stimulus is a proposed 8 million-square-foot mixed-use project for which CN Real Estate has received approval after years of negotiation. Located on an 85-acre downtown site near the waterfront, which is part of a 100-acre cleared parcel near the CN Tower and Skydome, the project involves 4 million square feet of commercial space and 4 million square feet of residential space. It will likely start modestly, while the office market continues to recover and the housing market remains sluggish. Another new development in the financial core, the 760,000-square-foot One Simcoe Place, is under construction. It will be the new home of the Workers Compensation Board and will add another 230,000 square feet of rentable space to the market.

The downtown retail market is thriving. However, U.S.–based big-box retailers, such as Wal-Mart and Home Depot, view the region as a prime target.

The Golden Age of Transit

In the 1950s and 1960s, Canada lagged behind the United States in freeway building. When Canadian citizens observed the negative impacts of freeways on U.S. downtowns and inner-city neighborhoods, they mobilized and stopped freeways—not only in Toronto, but also in Vancouver, Edmonton, Winnipeg, and Montreal. There was renewed interest in transit throughout Canada. The 1970s were the golden age of transit in Canada, as provincial governments provided generous capital and operating funds. Light-rail lines were opened in Edmonton in 1978 and Calgary in 1981, exclusive busways were developed in Ottawa and Winnipeg, and the Vancouver skytrain

was a feature of that city's World's Fair in 1986. Both Montreal and Toronto embarked on a second phase of rapid-rail transit construction in the 1970s.

In Toronto, the opposition to the freeway components of the 1966 transportation plan provided further support for the transit elements and, according to educator and researcher Richard Soberman, "resulted in strong biases toward capital intensive transit solutions that remain to this day."[21] In 1971, the provincial government halted construction of the Spadina expressway by refusing to pay for its half share of the costs. Instead, it increased funding for mass transit.[22] In a 1971 speech, Ontario Premier William Davis famously declared that "cities are for people, not for cars." Shortly thereafter, he announced that Ontario's share of capital spending for transit projects would increase to 75 percent (compared with 50 percent for roads) and that transit operating subsidies would increase and a government-owned corporation, the Urban Transportation Development Corporation, would be created to develop new transit technology. Between 1973 and 1980, 14 miles of new subways were built.

Commuter rail had been initiated in 1965 to fill a gap in transit services—connections between Metro Toronto and the satellite suburbs not served by the TTC. It was beginning to become a factor in Toronto's travel picture. Between 1970 and 1985, GO (Government of Ontario) train ridership increased by 300 percent.[23]

One fallout from the cancellation of the Spadina expressway was an agreement in 1972 between the provincial and Metro governments to jointly prepare a new transportation plan, and this time to include the public, whose antagonism had doomed the 1966 plan. The Metropolitan Toronto Transportation Plan Review (MTTPR), modeled after a similar effort in Boston that also was undertaken after the defeat of a freeway plan, considered a range of transportation and land use options. It produced a policy plan (not a detailed plan) that established a foundation for transit-oriented regional subcenters. According to Richard Soberman, who served as project director for this review, it also freed transit managers from their tradition of using cost-effectiveness as the basis for recommending transportation improvements—in favor of using land use benefits as a rationale for subway construction.[24] This change in attitude is hinted at in a 1985 report reviewing transit proposals. "Is rapid transit cost-effective?" asks the report, and then answers that "it is impossible to prove conclusively, because one would have to imagine a future without subways. Ultimately, rapid transit is a decision about economic development, quality of life, and urban form, and so far, the track record has been a positive one."[25]

The Center Cannot Hold: The 1980 Plan

In the 1970s, a move to limit downtown growth coincided with a new concept of metropolitan form—multiple, decentralized urban centers instead of a single downtown—that became the basis for a regional plan that was adopted in 1980.

In Toronto, as in other Canadian cities but unlike many large U.S. cities, central city and suburbs enjoy a relative equality in terms of average incomes. Income tax data for 1989 showed an average household income for the city of Toronto of $38,000, slightly higher than the average income for Metro Toronto and for the Greater Toronto Area, both of which had average household incomes slightly over $36,000.[26] While southern Ontario's industrial heartland shared some of the economic problems of the U.S. Midwest during the 1970s, its cities did not contain a spatially segregated underclass.[27] The metropolitan government structure has helped prevent the problem of poor city versus fortified suburb that confronts cities such as Detroit and Cleveland.

When large-scale urban renewal was the vogue, Toronto undertook some megaprojects, but it quickly changed the focus of its renewal efforts to preservation.[28] In the 1970s, the central city was reasonably affluent, and it had stable neighborhoods and an active university and professional community. In 1972, a group of "reform" aldermen gained control of the city council and stopped downtown development. They did not, however, oppose the continuation of transit construction in the Spadina corridor, where the expressway had been stopped.

Metro's 1976 central area plan encouraged the deconcentration of office space throughout the region by establishing limits on office development in the central area and stopping all downtown transportation improvements for a ten- to 15-year period. Planners were afraid that more downtown development would put too much strain on the existing transportation network. They made assumptions about the number of additional daily trips that could be added during peak hours on roads into downtown (none) and transit (between 38,700 and 46,700), about the peaking of work trips, and about the potential for walking to work. Thus they backed into an estimate of the amount of additional office floor space that could be accommodated within the existing transportation network. It came to between 2.5 million and 4 million square feet.[29]

A regional pattern of subcenters of jobs and housing would allow more two-way use of transportation facilities during peak periods and help the city in its move to limit downtown growth and preserve the character of downtown. It would also provide functional downtowns for suburban boroughs—a strong selling point—and reduce average trip lengths by distributing jobs more evenly. In many ways, the subcenters concept certified existing trends. The subcenters concept was admittedly decentralizing, but it could give focus to the suburbanization that was naturally occurring and minimize its automobile orientation.

The regional plan adopted by Metro Toronto in 1980 identified three types of urban centers:

- **Downtown Toronto** would remain the dominant central area, and the primary location for government, corporate headquarters, and financial institutions. It would maintain its preeminent position in the artistic and cultural life of metropolitan Toronto.
- **Two major centers**—the North York center and Scarborough Town Centre—would emerge. Each would accommodate 30,000 to 40,000 jobs and include a mix of government services, cultural facilities, and entertainment services of the kind then found only in the downtown.
- **Four intermediate centers** with 5,000 to 10,000 jobs would be located on transit lines, and they would attract new growth that "might otherwise be dispersed or located in the downtown."

All centers were planned for higher densities than prevailed in their surrounding areas and they would have well-defined boundaries to prevent commercial encroachment on residential neighborhoods. For the Scarborough Town Centre, development was to be encouraged by a transit operating subsidy from Ontario that would be linked to commitments by the municipality to adopt land use policies that would enhance the use of the subway line serving it.

The special LRT (light-rail transit) subsidy—offered in recognition of the chicken and egg problem of trying to build transit to encourage development in areas that currently lack transit ridership because there is no development yet—was originally estimated at $13 million (in 1979 dollars) over a ten-year period. (Under the provincial formula then in place, the LRT would have been limited to a special subsidy of $2.2 million over three years.) The deal contained both an assurance portion, which declined as ridership grew, and an incentive portion tied to the success of the Town Centre. The borough of Scarborough agreed to adhere to official plans supporting development that is oriented to the transit station, to adopt appropriate development plans and zoning, and to complete a Town Centre plan before the LRT line opened.[30]

115

Agreement was reached on the land use element of this transit-oriented plan in the late 1970s. Somewhat ironically, one of the first transportation actions of the plan was the elimination of a subway line that would have served a new radial corridor to downtown. This was because the dispersion of jobs away from downtown was expected to preclude the need for more downtown rail transit. By the time the plan was finally adopted in 1980, all the other subways planned to support the grid of transit-oriented centers also had been eliminated.[31] The 1980 plan included few major new transit facilities. There appeared to be concern that proposed transit investments could disrupt communities and, in particular, detract from their redevelopment potential.

Looking forward to 1980, the original 1966 transportation plan had recommended 105 miles of expressways. Despite the highly publicized conflict over the Spadina expressway, 80 miles of expressways actually had been built by 1980, although none of these were in the city of Toronto. Of the 37 miles of rapid-rail transit recommended, 39 miles had been completed and four miles were in the works, for a total of 41 miles. Overall, not a bad record—76 percent of the highways that were originally proposed and more than 100 percent of the transit had been put in place. Even more impressive, the transit service level in 1980 in terms of miles of transit was more than triple the 1954 level. Per capita ridership reached 201 annual trips by 1980—a 23 percent gain since 1966.[32]

The Location of Jobs and Travel

Auto ownership in Metro Toronto is low and growing only slowly, from less than one car per household in 1964, to one in 1980 and 1.2 in 1986 and also in 1991. Rapid population growth in the auto-oriented outer suburbs has meant higher rates of auto ownership for the larger region—the Greater Toronto Area. In 1964, auto ownership in Greater Toronto was also less than one car per household, but by 1986 this had grown to 1.4 cars per household.[33] (In contrast, U.S. auto ownership averaged 1.1 vehicle per household in 1969 and increased rapidly to 1.6 cars per household in 1977. New York ranks lowest in vehicle ownership among large U.S. metropolitan areas. That region's 1990 average of 1.2 vehicles per household was equal to Metro Toronto's in 1986.[34])

The transportation impact of population and job shifts has been mixed. The location of many of the region's new jobs in urban areas served by transit and the gentrification of close-in neighborhoods have reinforced the transit market. The movement of population to distant suburbs has increased driving, but the movement of jobs to the suburbs has decreased

some driving distances. The length of the average commute trip increased from six miles in 1971 to 7.5 miles in 1981. In the outer suburbs, the average commute distance actually decreased during this period because of the wider availability of jobs. Toronto's new development patterns have shifted the focus of transportation demand from the CBD to the circumference, challenging planners who are most comfortable developing strategies to get people downtown.[35]

In the Toronto central area, the amount of office space tripled from 20 million square feet in 1960 to 62 million square feet in 1985. Annual additions to the office inventory jumped from 5 percent during the 1960s to 10 percent in the early 1970s, the period when concern over excessive downtown development was at its peak. Following the adoption of the 1976 central area plan, office development slowed, although it increased again in the early 1980s—with office space growing 18 percent between 1980 and 1984.

Downtown office employment grew more slowly. Employment expanded at an annual rate of 2 percent between 1960 and 1975, rising from 100,000 to 150,000. In the latter half of the 1970s until the recession of the early 1980s, employment grew at a faster 6 percent annual rate. The primary reason for the growth rate disparity for office space and office employment was a significant increase in the average amount of space per worker, from 207 square feet in 1960 to 280 square feet in 1985.[36]

The number of passenger trips into the central area has not increased at a level comparable to the increases in office floor space and office employment. Several factors appear to be involved. Population in the central area was growing, so more of the new jobs could be filled by people living inside the area, many of whom could walk to work. The central area was becoming an "executive" city with more white-collar jobs and fewer manufacturing jobs, which meant a less structured workday and less peaking of travel.

As shown in Figure 8-4, the number of persons entering the central area during the 7 a.m. to 9 a.m. peak period grew only slightly from about 230,000 in the mid-1970s to about 255,000 in the mid-1980s. While the number of autos remained virtually unchanged, possibly because of a deliberate policy not to expand highway capacity, there were significant shifts in modes of travel. The number of persons arriving by auto declined because of a drop in auto occupancy—from 1.53 persons per vehicle in 1960 to about 1.3 persons in 1970 and later. The number of persons arriving by transit increased significantly, from 52 percent in 1966 to 66 percent in 1985. There

Figure 8-4

People Entering the Toronto Central Area During Morning Peak Period by Travel Mode, 1960–1985[1]

			People Arriving by . . .					
	Streetcars[2]	Subway[2]	TTC Buses	Other Buses	Cars	GO Trains	Total Arrivals	Number of Cars
1985	13,970	120,147	6,831	3,156	87,700	24,185	255,989	66,698
1984	13,147	110,195	8,637	1,580	97,040	22,644	254,043	73,432
1983	15,302	116,477	9,146	2,098	93,857	20,730	257,610	70,846
1982	14,853	117,602	8,656	1,436	87,615	21,864	252,026	65,587
1981	15,485	114,325	8,117	1,648	89,541	18,951	248,067	67,500
1980	15,677	110,205	9,138	2,296	96,161	16,470	249,947	72,353
1979	15,438	107,718	7,918	4,364	92,589	15,446	243,473	69,694
1978	16,776	102,228	8,899	2,964	88,520	13,092	232,479	67,133
1977	17,092	97,663	12,687	3,699	92,255	11,579	234,975	69,405
1976	17,743	94,114	13,076	1,683	86,704	7,367	220,687	64,327
1975	17,468	98,618	15,678	2,884	86,999	9,041	230,688	65,172
1974	18,345	96,225	13,615	2,161	97,824	8,133	236,303	73,540
1973	18,687	80,586	13,528	2,109	87,584	7,494	209,988	65,746
1972	20,315	85,212	14,398	2,054	95,042	7,283	224,304	71,751
1971	19,502	81,768	13,741	1,223	98,291	6,700	221,225	73,720
1970	20,657	80,086	14,571	1,343	105,248	6,000	227,905	70,215
1969	53,184	45,360	16,860	1,310	103,431	0	220,145	68,925
1968	93,614	0	17,444	1,737	103,813	0	216,608	69,183
1967	98,600	0	13,170	1,461	97,064	0	210,295	63,810
1966	89,402	0	11,451	1,497	101,139	0	213,317	67,136
1965	94,510	0	5,784	1,159	97,166	0	198,619	65,166
1964	93,801	0	11,842	0	106,680	0	212,323	73,069
1963	94,121	0	12,442	0	112,037	0	218,600	74,758
1962	89,956	0	11,339	0	107,998	0	209,293	70,702
1961	93,140	0	8,176	0	103,466	0	205,158	68,542
1960	95,676	0	7,539	0	96,979	0	200,602	63,239

[1] 7 a.m. to 9 a.m. on survey day.
[2] Pre-1970 streetcar figures include subway passengers.
Source: The Transportation/Land Use Relationship within the Toronto Central Area, 1960–1985 (Stephen Woodward, 1989).

were shifts in transit modes. Fewer people arrived by streetcar and bus, and more arrived by subway and the GO commuter trains. GO train ridership increased by 300 percent between 1970 and 1985, as the rail network expanded. This shift to more congested transit modes—the subway and GO trains—created a false perception that transportation services to the central area were being pushed to the limit.[37]

Between 1985 and 1995, the number of persons entering the central area peaked at 324,000 in 1989, and then declined to 296,000, which was slightly below the 1985 level. Peak period (which was now reported for the three hours from 7 a.m. to 10 a.m.) vehicular traffic grew by 8 percent from 1985 to 1987, and then declined as the recession took hold; by 1995, it had regained its 1985 level. Combined transit ridership peaked at 212,000 trips in 1989, and then declined to 183,000 trips. While the 1995 transit share of 62 percent was the same as its share in 1985, TTC's combined subway, bus, and streetcar share had fallen to 50 percent. A significant jump in the GO commuter rail share to 12 percent made up the difference.[38]

By the mid-1980s, a number of important employment centers existed outside the central area

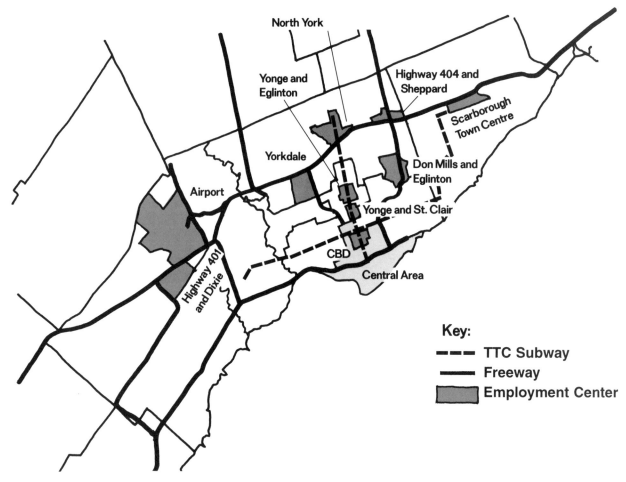

Major employment centers in the Greater Toronto Area, 1986.

Sarah Wells

(see map above). While some of these were the transit-oriented centers referred to in the 1980 plan, the two largest were the highway-oriented center near the airport and the Don Mills and Eglinton center, which also was primarily highway-oriented. These two centers attracted 55,000 daily work trips in 1986.[39]

From 1976 to 1984, employment growth around the two major centers in the subcenters plan was 42 percent, a growth rate that was exceeded by auto-oriented office parks, at which employment grew by 74 percent. By 1985, employment downtown was at 371,000. At the two major centers combined it was 36,000, and at the four intermediate centers combined it was 45,000—for a total employment of 81,000 at the six transit-oriented centers outside the central area. Total employment at the six auto-oriented office parks was less.

The two major centers in the plan—North York (which is located at Yonge & Sheppard) and Scarborough Town Centre—had established a respectable size by the mid 1980s. In 1986, the transit share for commuters to North York was 35 percent and for commuters to Scarborough Town Centre it was 22

percent—quite respectable transit shares for centers outside downtown.

New Transit Initiatives

When new transit projects were scaled back in the 1980 plan, the intention had been to revisit the question of rapid-rail transit extensions in five years. The review was accelerated for three reasons. First, transit ridership to downtown had grown significantly, causing crowding problems and calling into question the decision not to build any new radial transit lines. Second, if rapid-rail construction were stopped, the transit engineering department would have to be dismantled, a step not to be taken hastily. Third, the multicentered land use plan might not work without the transit grid. An external incentive came from Ontario's interest in promoting its own automated rail technology. Moreover, economic growth had slowed, and economic development had replaced growth management as a major concern. The 1979 OPEC oil embargo and subsequent runups in gasoline prices added further immediacy to the reconsideration of transportation plans.

With the prospects good for provincial funding of new transit projects, the TTC took the initiative in preparing an aggressive transit agenda. Its 1985 report, *Network 2011*, proposed five priority projects, phased over a 28-year period. It proposed spending $95 million to $100 million annually on subway construction, for a total of $2.7 billion through 2011. Improving suburban transit was seen as the first priority. Among the projects proposed were a (suburban) circumferential rapid-rail loop, a downtown "relief" subway line, and a busway in the Eglinton corridor. Even with minimal growth in overall transit ridership, the TTC calculated that the additional revenues from these services would greatly exceed the increased operating costs (compared with a conventional bus system), so no fare increase would be needed.[40]

A critical assumption in the affordability calculation was that the province of Ontario would pay 75 percent of the construction costs of the new transit lines, and Metro Toronto would pay 25 percent. Outlying municipalities protested. They were concerned that an expenditure of that magnitude would preempt a planned highway that was needed to serve the phenomenal growth that they had been experiencing.

Outlook: A New Plan, a New Government, and Unresolved Problems of Governance And Growth

Transit ridership has been falling since the late 1980s as a consequence of job losses in downtown and, possibly, continued suburbanization. Despite an abundance of transit proposals in the 1970s and 1980s, only some minor extensions were accomplished. Moreover, regional governance, once Toronto's strong point, has been compromised and is a major issue.

Reurbanizing Toronto

Metro Toronto adopted a new plan in 1994, the theme of which is reurbanization. The idea is to capture a good share—if not its traditional share—of the regional growth that will occur. Reurbanization in a basically built-up area with little land available is a difficult challenge.

A fundamental goal of the new plan is that all new development be served by transit in order to reduce the negative transportation impacts of growth. The earlier multicentered approach has been expanded with the addition of corridors and infill areas. A third major center—Etobicoke—has been added to the west, in the direction of most of the residential growth out-

side Metro Toronto. Size and density targets have been set for these centers, but they were not incorporated formally into the plan. Urban design guidelines and parking standards are also suggested.

Looking toward 2011, the plan targets an additional 200,000 to 245,000 housing units. Job targets for employment centers are shown in Figure 8-5. In a turnaround from previous planning assumptions, employment in the central area is expected to grow substantially, from 385,000 to 560,000. Jobs at the new major center in Etobicoke are planned to increase from 8,000 in 1992 to 25,000 at the center's maturity. Employment at Scarborough Town Centre and North York is planned to increase to 50,000 and 65,000, respectively. Intermediate centers are planned to also capture a significant share of employment gains. A key factor in this plan, as it was in the 1980 plan, is the rapid-rail transit network.

Waning Provincial Support for Transit

Provincial elections in 1995 brought a different party to power, which took a much more conservative approach toward taxes, spending, and government regulations. In 1990, Ontario had agreed to support the construction of four new rail-transit lines in Toronto—setting off, as noted, years of controversy on the local level. In 1995, Ontario called the deal off.

When these transit lines were being planned, the prospect of financing a significant share of local costs with assessments on development had been discussed. The development community had expressed some interest, but the uncertainty of provincial government backing and the weak economy prevented the working out of any definite arrangements. In addition, as Robert Pringle, a transportation planner for Metro Toronto points out, "it seemed counterproductive to charge developers for building where we wanted." When the new provincial government took office, the Eglinton extension, which was under construction, was canceled, to the outrage of, among others, the mayor of York, Francis Nunziata, who pointed out that private developers had made commitments related to the subway project and that canceling it would make them even more wary than they were of entering into partnerships with the government—at the same time that the government was committed to stimulating private sector investment. He also urged Ontario to reconsider this decision in light of the damage it would do to Metro Toronto's reurbanization plans.[41]

Construction on the Sheppard Avenue subway, originally proposed for private support, began in 1995 with government funding, but was suspended in 1996, when Ontario announced a cut of $117 million from its share of the $930 million project. The province

Figure 8-5

Growth Targets for Toronto Employment Centers, 2011

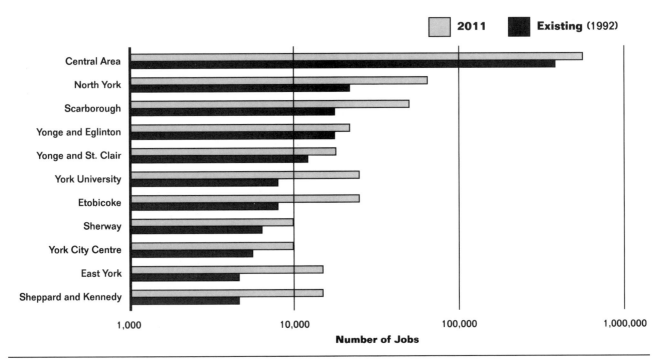

Source: Municipality of Metropolitan Toronto Planning Department.

is looking to the TTC to put forward a more cost-effective design and to Metro to carry a greater share of the cost. Both the TTC and Metro balked at the additional burden, which would probably have killed the project. In June of 1996, however, the mayor of North York proposed that the city could make up $90 million of the difference through development fees.[42]

The chief administrative officer of the Metro Council initially recommended mothballing the project, but Metro then committed to paying for tunneling, without stations or track work, subject to a combination of local contributions and reduced construction costs. TTC was requested to downgrade its Cadillac design to a Chevrolet.

Regional Governance

The two-tier form of metropolitan government that was adopted in 1953 responded to two problems: how to finance suburban growth and how to save the inner city from overcrowding, physical deterioration, and immobility. At the same time, it allowed local elected officials to remain in office, an important political consideration. This model rejected the municipal self-reliance idea on which many U.S. cities with the same problems relied. The model of

cities operating independently in a region was, the planners thought, a recipe for creating fierce competition for jobs and tax base, municipal winners and losers, and many urban problems.

The Metro government deals with regional concerns while the municipalities attend to local concerns. Areawide services are provided out of a pool of tax revenues collected from municipalities. The original federation consisted of the city of Toronto and 12 suburban, mostly rural, communities. Subsequent annexations have reduced the number of members to six: Toronto, North York, Scarborough, Etobicoke, York, and East York.

The Metro Council provides regional infrastructure and social services, including public transit, major roads, water and sewers, recycling, police, welfare, childcare, ambulance services, and regional parks.

Local government responsibilities include residential streets, zoning, fire protection, local libraries, local parks, public health, and garbage collection.

Planning and economic development are shared responsibilities. Metro was initially given planning authority over 720 square miles, considerably more territory—240 square miles—than its constituents occupy. When Ontario established the four surrounding regions—Durham, York, Peel, and Halton—Metro's territorial jurisdiction was scaled back.

This two-tier system has made for effective government in Toronto. Its big limitation today is that the Toronto region has grown well beyond the geographic boundaries of the Metro government. (The four surrounding regions also have metropolitan governments, which were created by the provincial government in the 1970s.) Traffic in greater Toronto does not respect regional boundaries. In 1993, the number of private vehicles entering Metro Toronto daily between 6:30 a.m. and 6:30 p.m. was 1.4 million—a 55 percent jump from 1983. A not insignificant number of persons entering Metro Toronto arrives by transit—10 percent. Still, the vast majority of inbound travelers comes in cars, signifying that the Toronto suburbs are becoming the type of auto-oriented metropolitan region that many have worked hard to prevent.

People in the region generally agree on what the needs are. Pressing needs include the reform of regional property tax assessments; the planning and coordination of regional services; governmental streamlining and clarification of responsibilities in order to disentangle decision making and ease the feeling that many have of being overgoverned; the preservation of a strong two-tier system of government; and higher levels of social services.

Not surprisingly, they are less in agreement on what to do. The GTA Task Force, a task force on governance in the Greater Toronto Area, was established before the latest provincial government change, with three main goals: reduce duplication, decrease complexity, and improve efficiency. It submitted a report to the new government in January 1996. The report calls for replacing the five regional governments with a weak Greater Toronto Council (GTC) consisting of mayors and other local elected officials; and for creating new quasi-governmental bodies called flexible service districts.

Metro Toronto estimates that under this proposal the number of governing bodies delivering ambulance services, licensing, water, sewage, arterial roads, and community services would increase by 600 percent in Metro Toronto alone. Its own recommendation is that existing regional governments should not be dismantled unless they are replaced with something just as effective.[43]

In this debate over the region's future, the city of York stresses the principle of areawide responsibility. Because it is a primary receiving area for immigrants and refugees, York has a disproportionate share of low-income residents. The city supports the two-tier system, or a replacement system that also spreads costs based on the ability to pay.[44] In the city's view, the first principle for evaluating governance options is that shared contributions and shared benefits require a sharing of local resources. Edward Sajecki, municipal commissioner of planning and economic development, points out that assuring a good basic education regardless of residence is important, because a student from anywhere can "become a contributing member of society anywhere in the greater region."

The option of expanding the boundaries of Metro Toronto to include the entire Greater Toronto Area was foreclosed when the other regional governments were created. The four other regions are mostly rural areas with some urban development at their fringes, and they are not likely to welcome being taken over by an urban area. As late as the mid-1990s, the top elected official in three outlying regions was a farmer. The option of carving off just the urban portions of these regions for incorporation into Metro Toronto region is politically sensitive at the provincial level. The expanded Metro Toronto would contain 40 percent of the province's population, and thus pose a potential challenge to provincial authority.

Financing Suburban Growth

Concern about the excessive costs of serving current patterns of growth is growing. The GTA Task Force commissioned a study on the economics of urban form, which attempted to address some tough questions: Are certain development patterns more cost-effective than others? What factors of urban form affect cost? Are the savings significant? Is low-density development at the fringe being subsidized? Who benefits from more efficient development patterns? (The study was carefully referred to as a study of "urban form" to avoid the pejorative "sprawl.")

The study estimated costs for three hypothetical scenarios: spread-out development (a continuation of current trends), centralized development (primarily infill growth), and nodal development (decentralized but clustered growth). If current development trends continued, an estimated 350 square miles of greenfield land would be urbanized in the next 25 years, an area equivalent to one and one-half times the current size of Metro Toronto. Centralized development would consume less than half that amount of new land and nodal development would use one-third less.

Over the next 25 years, said the urban form study, investment in infrastructure would total $90 billion if development continues to be spread out. This could be lowered by $10 billion to $16 billion under a more focused development future, which would also reduce maintenance and operations costs by $2.5 billion to $4 billion. The development of more flexible, cost-effective standards for infrastructure could produce further savings.

The four major categories of costs were transportation, other hard services, environmental services, and human services. Human services—which includes hospitals, social and other health services, schools, police and fire, recreation, and parks—represented the largest cost (around $30 billion for the spread-out development option), and the cost of providing human services varied little over the three alternatives.

That the transportation costs were also virtually the same—$27 billion to $28 billion—for all three scenarios is perhaps more surprising. The reason is that savings in road costs under the centralized and nodal scenarios were compensated by increases in transit investments, which were described as "massive." Also, the same basic network of arterial roads is needed when development is nodal as when it is spread out, a phenomenon referred to as the high costs of a discontinuous urban growth pattern. The study found, interestingly, that the road infrastructure savings of the nodal pattern compared with the spread-out pattern were greater for local roads, which are usually paid by developers, than for major roads and expressways.

Addressing another sensitive issue, the costs of urban form study found that the residents of higher-density urbanized areas effectively subsidize the residents of low-density areas under current mechanisms—frontage fees, property taxes, and income taxes—for financing infrastructure. Such subsidies distort housing and property markets. Cost-based pricing mechanisms like user fees and development charges could correct these distortions, and, the study concluded, such a market-oriented approach is preferable to trying to dictate urban form through planning and regulatory restrictions. To work, however, such pricing mechanisms should be consistently structured across the region.[45]

Shifting infrastructure costs to development could also help reduce the provincial government deficit. However, even with developer contributions, provincial assistance is likely to continue to be required for suburban development. According to Dale Martin, provincial facilitator for the Ontario Ministry of Municipal Affairs, with all the evidence that growth does not pay for itself and that new municipalities are unwilling to subsidize it, "it is hard to sustain the business case for suburban development." Perhaps the aversion of Ontario's government to taxes will force communities to internalize the costs of growth, and perhaps they will thus end up supporting compact development. The denial of provincial funds would cause suburban governments to turn down growth in undeveloped areas and lend support to Toronto's reurbanization policy. It would, in effect, create a fiscal growth boundary.

Lessons

Toronto wrote the textbook on creating a livable city and developing supportive infrastructure. But, it is a textbook that is becoming dated. It stands in need of revision.

Toronto's early success in encouraging development around transit involved good planning, but also good fortune. The subway was built during a period of growth. Supported by an influx of baby boomers into the city, permissive zoning near transit, and restrictions elsewhere, apartment developers found that around the subway was an obvious place to develop.

Transit's subsequent history has been less blessed by good fortune. On-again, off-again support from the provincial government has been a large part of the problem. Another part may be that the transit agency has lost its commitment to cost-efficiency. The early subway routes were built by the city of Toronto using its own funds. Later, transit managers began to embrace "strategic land use benefits as justification for subway expansion in a number of marginal situations," and they no longer argued "the need for subway expansion based on capacity requirements."[46] TTC brought on a new general manager in 1995, who has emphasized a return to efficient transit operations and the replacement of outdated equipment.

Local government support via zoning and regulatory approvals for higher-density development near transit has become unpredictable. A major rationale for the 1980 multicenter plan was to keep new development out of the central area and surrounding neighborhoods. Transit planners and developers can no longer count on municipal officials to permit densities that will support transit, unless they obtain such commitments in advance. (And sometimes even advance commitments are not honored.)

The diminished effectiveness of metropolitan government is a key issue. As former Toronto Mayor David Crombie said in 1994: "Toronto found a way to use Metro as an instrument of equality. That's why our core is not the hole in the doughnut. Before 1970, the Metro planning boundary was three times larger than now. Today, we seem to have lost the sense of being in an economic region." Metro's dominance within the greater region is being eroded. Tax differentials are a part of the problem, with many businesses in Metro Toronto feeling overly burdened by taxes and looking to outlying locations for tax relief.

Toronto has contributed many cutting-edge ideas and provided models for other metropolitan regions. Recently, however, U.S. transit operators and cities have been much more entrepreneurial in fostering development around transit. For Toronto, current budget concerns at the provincial level may reduce subsidies for suburban development and exert strong pressure for the reurbanization of the central area.

However, it appears that in Toronto, as in most North American urbanized areas, housing markets will continue to be dominated by single-family homes in the suburbs. Elizabeth McLaren, an assistant deputy minister in the Ontario Ministry of Municipal Affairs and Housing, with responsibility for developing a coordinated plan for the greater Toronto region, emphasizes a need "to plan the suburbs better, not to eliminate them." McLaren faults gold-plated standards—unnecessarily wide roads for fire trucks or excessively large land dedications for schools—as part of the problem. Suburban developers in Toronto can expect to lose half of a 200-acre subdivision to government land dedications.

In a search for new development models, some Toronto municipalities have brought in Florida-based architect Andres Duany, one of the leaders of the new urbanism movement, for advice. Whether bike paths will succeed in six-foot snow drifts remains to be seen.

Toronto is likely up to the challenge. Despite its problems, the region is viewed as one of the most desirable places in the world to live and to do business. Greater Toronto has an excellent planning tradition. Government is held in respect to a much greater degree than in most U.S. areas. When Toronto develops new ideas, it may become a new model for urban regions.

Notes

1. Richard M. Soberman, "Rethinking Urban Transportation: Lessons from Toronto" (paper to be presented at the annual meeting of the Transportation Research Board, January 1997).

2. According to the Corporate Resources Group, Geneva, Switzerland, which assessed cities for 42 quality-of-living factors under such categories as political and social environment, culture, health, education, public services, recreation, consumer goods, housing, and the natural environment, 1994.

3. According to PHH Fantus, based on a survey of 800 senior executives of U.S. multinational firms, 1993.

4. Larry McNally, "Roads, Streets, and Bridges," in Building Canada: A History of Public Works, ed. Norman R. Ball, for the Canadian Public Works Association (Toronto: University of Toronto Press, 1988), p. 32

5. Paul Andre-Linteau, "Urban Mass Transit," in Building Canada, pp. 60–69.

6. Toronto Transit Commission and Municipality of Metropolitan Toronto, Network 2011: A Rapid Transit Plan for Metropolitan Toronto (Toronto: Toronto Transit Commission, May 1985), p. 5.

7. McNally (see note 4), pp. 38, 44, 49.

8. Julie Harris, "Airports," in Building Canada, pp. 293, 306, 308.

9. IBI Group, Urban Travel and Sustainable Development: The Canadian Experience (prepared for Canada Mortgage and Housing Corporation, 1993), pp. 6–7.

10. Ibid., p. 8.

11. Juri Pill, "Emerging Suburban Activity Centers in Metro Toronto," Journal of the American Planning Association, summer 1984, pp. 310–315.

12. Juri Pill, Planning and Politics: The Metropolitan Toronto Transportation Plan Review (Cambridge, Massachusetts: MIT Press, 1979), p. 32.

13. Soberman.

14. Statistical comparisons based on Transportation Association of Canada, Urban Transportation Indicators in Eight Canadian Urban Areas (Ottawa: Transportation Association of Canada, 1996), appendix C, pp. 1–2.

15. Municipality of Metropolitan Toronto, Key Facts: 1992 (Toronto: Municipality of Metropolitan Toronto, 1992), p. 1-1.

16. Juri Pill, "Metro's Future: Vienna Surrounded by Phoenix?" Toronto Star, February 15, 1990.

17. Royal LePage Strategic Advisory Services, "Toronto Metropolitan Area," in ULI Market Profiles: 1995 (Washington, D.C.: ULI–the Urban Land Institute, 1995), p. 359.

18. David M. Nowlan, "The Changing Toronto Area Economy" (paper presented at the Forum on the Future of Industrial Land, sponsored by Metro Toronto, December 1994).

19. Key Facts: 1992, p. 4-1.

20. Alfred Holden, "Why Toronto Works," Planning, March 1995, pp. 2–10.

21. Soberman.

22. Frances Frisken, "The Contributions of Metropolitan Government to the Success of Toronto's Public Transit System: An Empirical Dissent from the Public-Choice Paradigm," Urban Affairs Quarterly, December 1991.

23. Stephen Woodward, The Transportation/Land Use Relationship within the Toronto Central Area: 1960–1985, University of Toronto Master's Program in Planning, Paper no. 34 (Toronto: University of Toronto, April 1989).

24. Soberman.

25. Network 2011 (see note 6), p. ii.

26. Patterns of immigration are changing and may affect income averages. Between 1980 and 1985, Great Britain and the United States—generally the source of relatively high-income immigrants—were the first and third highest sources of immigrants to Canada. In the following five years, they dropped to fifth and seventh place, behind Hong Kong, Poland, India, the Philippines, and Vietnam. See Key Facts: 1992, table 1-15 for immigration and table 2-19 for income.

27. Michael A. Goldberg and John Mercer, The Myth of the North American City: Continentalism Challenged (Van-

couver: University of British Colombia Press, 1986), pp. 210, 213.

28. Alfred Holden (see note 20).

29. Woodward, p. 2.

30. Municipality of Metropolitan Toronto, Report no. 13 of the Metropolitan Executive Committee, item 14, adopted May 23, 1980.

31. Pill, "Emerging Centers" (see note 11), pp. 306, 310.

32. Frisken (see note 22), p. 277.

33. City of Toronto, *Evaluating the Role of the Automobile: A Municipal Strategy* (Toronto: City of Toronto, 1991), p. 117.

34. Michael A. Rossetti and Barbara S. Eversole, *Journey-to-Work Trends in the United States and Its Major Metropolitan Areas, 1960–1990* (Washington, D.C.: U.S. Department of Transportation, 1993), pp. 6–7.

35. B. G. Hutchinson, "Structural Changes in Commuting in the Toronto Region: 1971–1981," *Transport Reviews*, 1986, v. 6, no. 4, pp. 317, 325.

36. Woodward, pp. 17–18.

37. Ibid., p.24.

38. Municipality of Metropolitan Toronto, *Metro Cordon Count—1995* (Toronto: Municipality of Metropolitan Toronto, July 1996), p. 6.

39. Sarah Wells, *Analysis of the Differential Impacts of Transit Modes on Travel Behavior in the Greater Toronto Area* University of Waterloo Ph.D. thesis (Waterloo, Ontario: University of Waterloo, 1994), p. 67.

40. *Network 2011*, p. xi.

41. Francis Nunziata (Mayor, City of York), letter to The Honourable Mike Harris (Premier of Ontario), July 13, 1995.

42. "Plan to Save Subway Line Proposed," *Globe and Mail* (Toronto), June 12, 1996.

43. Municipality of Metropolitan Toronto, *Metro's Plan for Reform* (Toronto: Municipality of Metropolitan Toronto, April 2, 1996).

44. City of York, *Managing the GTA as a Community: A Creative Response to the Challenge* (York, Ontario: City of York, 1995), p. 1.

45. Pamela Blais, *The Economics of Urban Form* (prepared for the GTA Task Force, December 1995), pp. 22, ii.

46. Soberman.

Chapter 9

Transportation and Development in San Diego
Protecting Paradise

San Diego is an auto city. . . . The rapid spread of the region is associated with the automobile, and so is the disruption of residential neighborhoods by freeways.

— Kevin Lynch and Donald Appleyard, *Temporary Paradise*, 1974

Many industry observers regard San Diego's light-rail transit as the most successful implementation of LRT in the United States since the end of World War II.

— John Kain, Harvard University, 1995

San Diego has long been a fast-growing region in a fast-growing state. It has developed a range of approaches for dealing with growth. Some of the earliest growth management programs in the country came out of San Diego, and the city is a pioneer in regional transportation planning and development. Among the notable programs and projects that San Diego has put in place to cope with growth are the following:

■ A city growth management program that promotes growth in intown neighborhoods;

■ The revitalization of downtown, including major retail development;

■ A new rail system constructed without federal funding;

■ The assignment of growth management responsibility to a voluntary regional government association; and

■ The establishment of a regionally managed transportation improvement program.

San Diego is one of only two large regions in the United States—the other is Phoenix—that ranked among the top five in population growth for both the 1970s and the 1980s. The transportation consequences of such extraordinary growth are predictable: inadequate facilities in the urbanizing areas and deteriorating facilities in the central areas. San Diego has been at the cutting edge in thinking about the relationship between development and transportation. The manner in which the region has encouraged inner-city revitalization, managed suburban growth, and provided a range of transportation services makes it a model for many other U.S. regions.

Before the 1970s: Early Transportation Planning

San Diego's first tourist was Juan Cabrillo, who sailed into the bay in 1542. Two centuries later, the first developer was Father Junipero Serra, who established the first in a chain of 21 missions in California. In 1769, Father Serra and Captain Gaspar de Portola left San Diego to find Monterey, and in the process broke a trail for what eventually became California's first road, el Camino Real—the King's Highway—which, in turn, became segments of California 1, U.S. 101, and I-5. To the east, a 1770 trail between the original presidio and mission was the forerunner of U.S. 80 and I-8. Other trails led to the bay, water supplies, gardens and pastures, hunting grounds, and building materials and they formed the growing city's transportation network.[1]

Early Development

San Diego became the capital of California under the Mexican flag following Mexico's revolt against Spain in 1825. By 1830, most settlers had moved away from the protection of the presidio to the area known as Old Town, just down the hill. In the 1880s, Alonzo Horton, a merchant from Wisconsin who had come to California to be a developer, purchased 905 acres of land, and began developing a downtown closer to the bay. Until the late 19th century, periodic declines in population threatened the town's very existence.

A direct railroad connection to the east was not completed until 1919. Outpaced by Los Angeles, San Diego never developed as an important port.[2] The real turning point came during World War I. The development of military facilities at Point Loma (now the submarine base) and the Navy Radio Station established San Diego's military importance. The Naval Training Center and Marine Corps Recruiting Depot opened in the 1920s. The city flexed its new muscles in 1915 by hosting the Panama-California International Exposition to celebrate the opening of the Panama Canal. The exposition site in Balboa Park is still an important cultural center. In World War II, the headquarters of the Pacific fleet was moved to San Diego following the Japanese raid on Pearl Harbor. In the postwar period, aerospace and technology industries arrived to enrich the region's economic base.

San Diego has a surprisingly long history of public transportation. The first streetcars began operation in 1886, first pulled by horses or mules and later powered by steam and cable. Eventually a network of electric streetcar lines wound throughout the city and interurban lines made longer runs into the countryside. Residential development, which had been clustered near the downtown and south along the bay, began to move inland toward El Cajon. The first independently operated motor buses appeared during World War I to serve outlying communities. San Diego Electric Railway Company eventually bought out many of these lines and began substituting buses for streetcars on its lightly used routes.[3]

Highway improvements in San Diego helped clear the way for the state's highway system. The first survey of county roads was completed in 1884, a year before the California legislature authorized a statewide survey of roads. The San Diego Road Commission was formed in 1908, and charged with building 1,250 miles of roads with the proceeds of a $2 million bond issue. In 1912, a group of San Diego businessmen provided money and lumber to build a plank road through the Imperial Valley sand dunes, using labor contributed by valley residents.[4]

Postwar Capacity Building

Wartime mobilization transformed the city from a sleepy border town to a booming metropolis. At Convair, one of San Diego's first aerospace manufacturers, Liberator bombers were manufactured around the clock and sometimes out-of-doors to take advantage of the favorable climate. The city's population grew 65 percent during the 1940s, and the region's population had reached 500,000 by 1950.

Wartime traffic congestion necessitated some demand-side measures, including the staggering of work hours at larger plants to reduce the number of cars on the road at any one time and priority parking for carpoolers. (This transportation strategy would return in the 1980s.) After the war, the transportation

Figure 9-1

Population of San Diego, 1970–1995

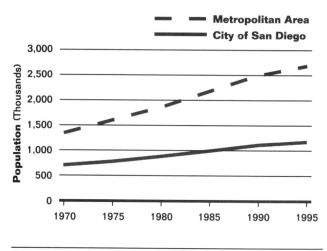

Sources: Decennial Census; and San Diego Association of Governments.

emphasis shifted from reducing demand to increasing the capacity of the highway system. At the same time, California was gearing up to implement its freeway system.

San Diego was an aggressive participant in freeway planning. The region launched one of the first cooperative transportation planning efforts in the country, dating from the mid-1950s. This effort resulted in a countywide circulation plan in 1967 and the creation of a formal council of governments at about the same time. Initially called the Comprehensive Planning Organization, the council of governments is now the San Diego Association of Governments (SANDAG).

Largely because of these early planning efforts, the region has four interstate highway routes, which the local office of the state highway department aggressively pursued because of the higher federal matching ratio for interstate facilities.

Suburban development followed the growing freeway system. Major regional shopping concentrations sprung up to the south, east, and north of downtown. Mission Valley was the largest of these developing retail areas. I-8 runs the length of the valley, which is crossed by the region's three other interstate routes as well as an older state freeway route. The development of this beautiful river valley into what is now a landscape of roads, malls, motels, and car dealers is often cited as an example of the excesses of the period. Mission Valley attracted a significant amount of office development as well, as did the planned industrial area of Kearny Mesa north of the valley. By the 1980s, retail and office development was spreading north along the I-5 and I-15 corridors.

At the same time, redevelopment began to occur in downtown (Centre City) San Diego and industrial development began in an agricultural area adjacent to the Mexican border 16 miles south of downtown.

The substitution of buses for streetcars continued, and the last of the old streetcar lines closed in 1949. The transit company had been sold in 1948 and reorganized as the San Diego Transit System. But declining ridership because of widening car ownership, and transit's growing inability to serve new auto-oriented suburbs sounded the death knell for private operation of the transit company. In 1967, the city of San Diego began operating the bus system as San Diego Transit Corporation. Several suburban communities started their own bus systems by buying equipment and contracting operations to private companies.

Current Conditions and Recent Initiatives: Guiding Growth and Providing Infrastructure

San Diego Compared with Other Metropolitan Regions

Compared with other large regions, San Diego ranked 11th in density in 1990, with 3,400 persons per square

mile. In their driving habit, San Diegans ranked eighth, with a daily average of 22 miles per capita. The region ranked even higher—sixth—in freeway miles, with an average of 0.75 miles of freeway lanes for every 1,000 people.

The region's drivers are highly dependent on these freeways. The freeway share of major road travel was 61 percent, putting San Diego second among large urban areas in the degree of freeway orientation, behind only San Francisco. Keeping up with traffic growth has been a problem. The number of miles traveled on the region's roads doubled during the 1980s, and San Diego became the sixth most severely congested large urban area in 1991 and the top metropolitan area in terms of the rate of increase in congestion.

Although transit accounted for a relatively low 3 percent share of commuting travel, San Diego ranked low—32nd—in solo drivers. One reason is that the region had the fifth highest level of walking and bicycling to work—4.6 percent. In 1980, when the region was much more compact, one out of every nine commuters walked or biked. As the U.S. and Mexican economies become more interrelated, the greater San Diego/Tijuana region is becoming a functional unit, and transit serves as an important means of travel between the two cities.

During the 1980s, San Diego ranked fifth in the growth of commuting by transit. It was one of only

Figure 9-2

Transportation Indicators for San Diego, 1970–1990

	1970	1980	1990	Percent Change 1970–1980	Percent Change 1980–1990
Population	1,357,854	1,861,846	2,498,016	37%	34%
Commuters	544,348	853,666	1,230,446	57	44
Private Vehicles	412,447	693,573	1,041,651	68	50
Driving Alone	365,288	611,093	950,262	67	56
Passengers	47,159	82,480	91,389	75	11
By Transit	13,069	27,816	39,376	113	42
Commute Share					
Transit	2%	3%	3%	36	–2
Private Vehicles	76%	81%	85%	7	4
Travel					
Daily Vehicle-Miles Traveled (Thousands)	n/a	25,451	50,286	n/a	98
Annual Transit Trips (Thousands)	13,111	30,193	51,493	130	71
Median Single-Family Lot Value	n/a	40,000	150,000	n/a	275

Sources: Decennial Census; *Journey-to-Work Trends in the United States and Its Major Metropolitan Areas, 1960–1990* (Federal Highway Adminstration, PL-94-012); and ULI Residential Land Price Surveys.

127

Figure 9-3

Population of San Diego by Distance from CBD, 1980 and 1990

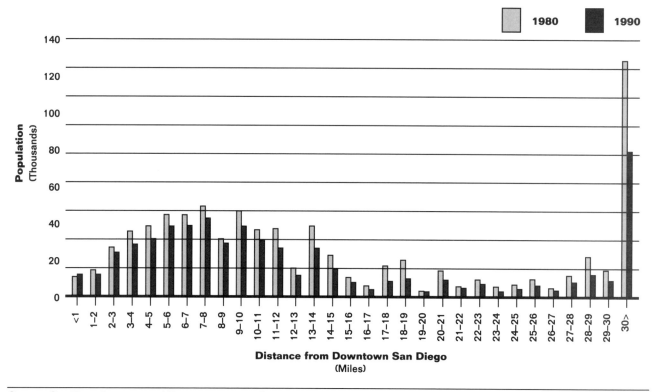

Sources: San Diego Association of Governments; 1980 Census; and 1990 Census

three regions not to lose transit share. Between 1980 and 1990, transit commuting increased by 41 percent and regional transit travel for all purposes grew by 70 percent, representing a 25 percent increase in per capita transit use, in sharp contrast to most other cities. Transit ridership gained another 27 percent by 1994. Concurrently, however, the number of commuters driving to work grew by 56 percent.

San Diego is, thus, a region with slightly above-average levels of driving and a heavy orientation of auto trips to freeways, combined with an above-average supply of freeways. Because of higher than average densities, the traffic is concentrated in a relatively small area. Strong gains in transit have allowed transit to maintain its market share for commuting. While transit provides important access to downtown and the Mexican border, its growth has not slowed down the amount of driving.

Population, Employment, and Development Markets

Growth in San Diego has historically occurred in cycles that often reflect military spending and the

state of the aerospace industry, the primary drivers of the economy. In the early 1970s, for example, the end of the Vietnam War combined with a national economic recession caused population growth to stall. The economy recovered and boomed again, until the recession of the early 1980s, which flattened the population growth rate until 1984. In the mid-1980s, population growth reached record levels.[5] Then the recession of the 1990s hit, the most severe downturn since the depression. The recession officially was declared over in 1994, when employment increased for the first time in four years.[6]

The region, which includes all incorporated and unincorporated portions of San Diego County, has a population over 2.6 million. The city of San Diego plays a strong role in the region. In 1994, the city accounted for 44 percent of the regional population (more people than lived in the other 17 cities combined), 45 percent of the retail sales, and 56 percent of the employment.[7]

Growth forecasts by SANDAG expect that between 1990 and 2015, the region's population will grow by 51 percent, housing units by 45 percent, and jobs by 30 percent. All these growth rate projections are considerably lower than past trends. The city of San Diego is anticipated to have slightly lower rates of growth: 42 percent for population, 37 percent for housing, and 23 percent for employment. The city remains a major magnet for growth, accounting for a 37 percent share of the region's housing growth and a 42 percent share of its employment growth.[8]

Most residential growth in the 1980s was located in areas 20 or more miles from the center (see Figure 9-3). This pattern is expected to continue and has major implications for the feasibility of transit, which attracts a significant share of its ridership from people living within about five miles of downtown (see Figure 9-4).

Housing in San Diego is among the most expensive in the United States. A ULI survey of land values suggests why: in 1990, the average residential lot cost $150,000, second only to San Jose among the 30 markets surveyed. That year, housing costs in San Diego as a percentage of household income were the highest in the nation.[9] Residential land values fell by one-third between 1990 and 1995, but they remain among the highest in the United States.[10] Much of the residential market is in the upscale North County or the more affordable East County. When the recovery

gets underway, it is expected that projects in South Bay will pick up, most prominent among them the 3,200-acre Eastlake development that started in the 1980s and Otay Ranch, a planned new community of 25,000 homes.

Downtown San Diego has seen a remarkable renaissance, and has become not only the region's primary office center, but also a significant retail and entertainment center with a growing residential base. During the 1980s, downtown office space exploded from 4 million to 11 million square feet. The redevelopment of downtown is discussed in the following section of this chapter.

Mission Valley, the region's first center of suburban activity, is a primary office and retail center. It contains over 5 million square feet of office space as well as Fashion Valley, San Diego's largest—and highest grossing—regional mall. Like downtown, Mission Valley has developed a substantial residential base. The Golden Triangle—an upscale center within the University City community near executive housing and south of the University of California at San Diego campus—has grown from virtually nothing in the early 1980s to a sizable center of office space (4 million square feet), retailing, and higher-density housing. Suburban edge cities have proliferated in

Figure 9-4

Transit Use by Commuters to Downtown San Diego by Distance Traveled, 1990

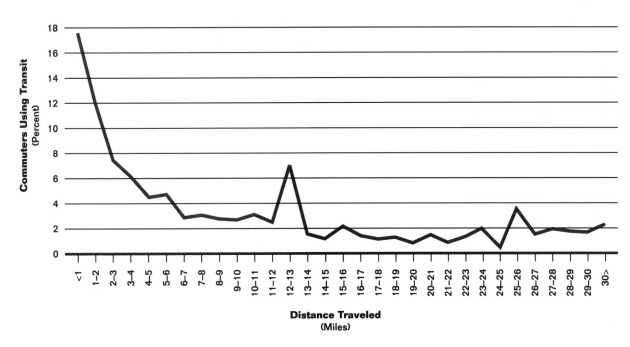

Sources: San Diego Association of Governments; and 1990 Census.

San Diego. According to one analyst: "This is a market of suburban business cores. While our downtown is important, it still employs only 10 percent of the work force in the region."[11]

Saving the Downtown

As the suburbs prospered, downtown San Diego deteriorated to a point in the late 1960s when tax revenues were not sufficient to cover basic city services such as police and fire protection. In 1972, newly elected Mayor Pete Wilson unveiled a two-part program to slow growth in the far fringes of the city and accommodate growth within developed areas. The mayor viewed the task of revitalizing downtown San Diego as very important and he promised to "do whatever we can to make our downtown livable, rather than a place from which people flee at day's end for the suburbs."[12]

A redevelopment effort modeled after Baltimore's was launched in 1976 with the creation of a nonprofit, public sector corporation, the Centre City Development Corporation (CCDC), to plan, implement, and direct redevelopment efforts in a 350-acre section of the blighted urban core. Its board of directors would be the sole negotiator between the city and private developers. Its primary source of revenue was tax increment financing. It had three major goals: draw retail business and residents to the area, create a strong job base and regional government center, and develop an effective public transportation system.

A plan had been adopted in 1975 calling for specific types of projects in four areas: large-scale retailing in the heart of downtown; a residential community in the Marina area; the extension of the city's traditional business district westward to the bay in the Columbia area; and a major convention center downtown. Revitalization of the Gaslamp Quarter, a Victorian era district that was listed on the National Register of Historic Places, was also planned.

The keystone was Horton Plaza, a project approved by the city council in 1972. A national request for proposals led to the selection of the Hahn Company of Los Angeles as developer in 1975. Eleven years later, Horton Plaza opened and became an immediate success with residents, planners and architectural critics, and shoppers. In 1995, the Hahn Company completed a major renovation in time for the tenth anniversary, which included a $16 million conversion of a former department store into a multiuse structure with a new Planet Hollywood restaurant, a two-level Sam Goody music store, and other national retailers. The complete Horton Plaza redevelopment project, of which Horton Plaza is the anchor, consists of 15 blocks of which all but a few parcels have been

redeveloped. It includes 237 housing units and 452 hotel rooms, as well as 2 million square feet of office and retail space. Between 1972 and 1994, assessed values rose by $415 million, representing a considerable return on the redevelopment agency's investment of $33 million for land acquisition and public improvements.[13]

Considerable office development also took place, more than doubling the amount of office space in downtown. The proposed convention center suffered a major setback when voters rejected a planned downtown location in 1981. It was eventually built on the waterfront, and has become a huge financial success and popular landmark, and plans for its expansion are in the works. The historic Gaslamp Quarter has become a major entertainment and restaurant venue. Housing development has helped round out downtown.

In 1992, CCDC's area of responsibility was expanded to almost 1,500 acres. The entire downtown became one big redevelopment project. By 1995, an estimated $2 billion had been invested in redevelopment projects in downtown, 90 percent of which was private funds. There were 75,000 jobs downtown and more than 15,000 residents.[14]

Many of the new developments support the expanded transit network. Two major projects—the MTS/Mills Building and One America Plaza—were joint endeavors of CCDC, the Metropolitan Transit Development Board, and private developers. Downtown's new resident population is served by a 40,000-square-foot Ralph's grocery store that opened in 1996 in the Marina area.

Guiding Growth

Following the adoption in 1970 of an ordinance that required adequate facilities concurrent with proposed developments, growth management began in earnest in 1974, when Mayor Wilson responded to picketing mothers in the new suburb of Mira Mesa where there were no schools and few services. In the same year, *Temporary Paradise*, a privately sponsored study that was intended to provoke discussion, was prepared by urban design experts Kevin Lynch and Donald Appleyard. This "illustrated discussion of the landscape of San Diego" called for development patterns that were more consistent with the natural setting. To residents, the visual appearance of development was one of the most objectionable aspects of growth. *Temporary Paradise* took a very broad view of the region, looking as far as Tijuana. In 1979, the city council adopted a growth management plan that was designed to prevent leapfrog development, finance growth through developer fees, and create an urban reserve for land on the periphery.[15]

The Tier Plan. The 1979 plan established four tiers of development within the city's 320 square miles: urbanized, planned urbanizing, future urbanizing, and parks and open space. (The parks and open space tier was later dropped as legally unsupportable.) The city proposed to encourage infill development in the urbanized areas by focusing its capital improvements there and charging no impact fees. In planned urbanizing areas, developers were required to provide improvements and pay impact fees. Future urbanizing areas were off-limits to development for 20 years.

Aided by the renewal of downtown and the rediscovery of city living by young, upwardly mobile residents, the plan succeeded beyond its authors' dreams. Two-thirds of the population increase over the first five years was located in the urbanized areas. Growth in the outer areas of the city amounted to less than two-thirds of original projections. In addition, the city council rebuffed efforts to reclassify land in the future urbanizing areas to allow development. The council's attitude was reinforced by a citizens initiative in 1985 that required voter approval of any reclassifications.

However, outward growth persisted. In an effort to bring it under control, the city council enacted an 18-month Interim Development Ordinance in 1987 that limited residential building permits to 8,000 per year, a level that was about half the demand that had been experienced in the preceding years. A citizens initiative to further cut back growth was defeated in 1988, after a high-powered campaign by development interests. However, growth limits were adopted also in 11 communities surrounding San Diego.

In recent years, developers have sought to develop in the urban reserve. Several have explored the feasibility of low-density golf course communities, because clustered development at an overall density of one unit per ten acres is allowed. Although some development of this type has occurred, most appears to be not feasible, and developers have sought rezoning. A ballot issue on the rezoning question failed to pass, however, leaving the urban reserve relatively untouched except for a few projects that have been allowed.

Through its Interim Development Ordinance, the city has stiffened environmental protection requirements for hillsides, canyons, floodplains, and wetlands. Residents have long treasured the region's recreational areas and open space, and interest in retaining these assets runs high. In fact, one of the stated purposes of restraining growth was to protect the quality of the community's open space.

The Unresolved Issue of Infrastructure Funding. In theory, the saving in infrastructure costs achieved by slowing down suburban development was to have provided funds for improving facilities in city neighborhoods. Two problems waylaid this effort to support close-in growth: state restrictions on taxation and the uncertain legal status of impact fees as a funding device.

In 1978, the voters of California approved Proposition 13, which rolled back property assessments to their 1975 market value and limited property taxes to 1 percent of property value. Adding to these restrictions, the Gann initiative a year later limited the growth of state and local spending based on inflation and population growth rates. Local revenues from property taxes plummeted. San Diego's policy of relying on general revenues to fund capital improvements in urbanized areas was rendered unworkable.

To respond to the issue of the legality of impact fees, in 1982 the city formulated a facilities benefit assessment program. Planners prepared capital improvements programs for designated areas, covering transportation, parks, fire protection, libraries, and other public facilities. The costs of these improvements were allocated to anticipated development and collected in the form of one-time fees as development occurred. The fees collected resided in a special fund until expended for programmed facilities. In effect, the program levied impact fees within special taxing districts. There are now more than 20 such districts.

The forced reliance on impact fees spelled trouble for the city's infrastructure program. First, the city's approach was litigated by development groups over several years. The litigation left plans for expending collected fees in limbo. When the city finally moved to design and contract for needed improvements, it proceeded slowly. The result was a backlog of $68 million sitting unused in the special fund while developing neighborhoods waited for promised facilities.

Second, the reliance on revenues from fees on development in urbanizing areas meant that the city had limited funds to invest in facilities in close-in neighborhoods, where a considerable amount of development was taking place. Also, a rise in standards in the 1970s and 1980s for most facilities, rendered many existing facilities in these neighborhoods obsolete. It was understandable, therefore, that residents became unhappy with the city's facility funding program and pressed for new solutions.

In 1987, after court decisions had validated its right to levy fees, the city began to impose impact fees on new development in urbanized areas, without establishing districts. By that time, residents of these areas were pressing for new solutions, among which was a series of restrictive regulations, including downzoning, that reduced opportunities for infill development.

Financing for public facilities remains problematic. Revenues from development fees dried up during the real estate recession and property reassessments reflecting the downward spiral of real estate in California have reduced revenues still more. Despite the city council's acknowledgment of the importance of the funding issue, a solution has not been found.[16]

The SANDAG Plan. It is a measure of San Diego's economic and political strength in the region that the city could actually adopt significant growth policies on its own. In 1988, a flurry of citizens ballot initiatives—in the home of ballot-box planning—mandating limits on growth at both the city and county levels all failed, partly because of San Diegan's traditional hands-off attitude toward growth and partly because of fears of an upcoming economic downturn.

Proposition C, an advisory measure, did pass. This established a regional board and called on the county and its 18 cities to resolve problems of transportation, air quality, solid waste, water and sewers, and industrial zoning.[17] In an unusual vote of confidence in a voluntary regional agency by its constituent governments, the responsibility for developing a plan was given to the San Diego Association of Governments in 1990.

Looking at the regional consequences of current development patterns in 1994, SANDAG concluded that the current plans of each city and the county together would severely exacerbate freeway congestion, widen regional inequities in access to jobs, and consume so much land that wildlife habitat would nearly disappear, along with land for residential development.

Increasing the density of residential development and focusing it around transit stations, SANDAG found, would reduce development pressure on rural land, open up perhaps 10,000 acres for habitat preservation, and offer modest improvements in freeway congestion and air quality. For example, 120 miles of freeway would experience severe congestion under the trends forecast, compared to 89 miles under the clustered alternative.

In 1995, SANDAG adopted a land use distribution element to help implement a clustered development policy through participating local governments. The policy has five components:

- Mixed-use development near transit and in community centers;
- Higher minimum development densities near transit;
- Lower development intensities away from transit;
- Multimodal transportation facilities; and
- Residential districts in major employment areas.

In its attempts to establish a consistent development policy among the region's governments, SANDAG has found that elected officials are more supportive of neotraditional, transit-oriented development concepts than are planning directors, who have a greater understanding of the political opposition that higher densities generate. Smaller cities in the region have not dealt with the specifics of new development policies.

SANDAG seeks to justify the higher-density, transit-oriented development that it is promoting with quantifiable measures, like reduced traffic congestion. However, because the improvements in congestion from such approaches appear to be relatively small, it may be easier to justify regional approaches to land use on the basis of a quality-of-life rationale, rather than simply as a means of reducing congestion.

Transportation Improvements

Revitalizing Transit. The regionalization of transit in San Diego began in 1975. Confusion over the proliferation of different bus operations in the region led the California legislature to create the Metropolitan Transit Development Board (MTDB), which has helped standardize and coordinate operations. MTDB is an organizational shell for a federation of six operating agencies: Chula Vista Transit, MTS Contract Services, National City Transit, San Diego Transit, San Diego Trolley, and the San Diego County Transit System. The operators share a common fare structure, route numbering system, and public information system, presenting a unified service to the rider but allowing great flexibility in the provision of service through private contractors as appropriate. MTDB also regulates taxi and jitney fares in the city of San Diego. Keeping operating costs down is a key issue for all the agencies. A similar transit agency serves the area north of the city of San Diego.

Locals consider the San Diego Trolley to be the crowning glory of the transit system. Authorized by its enabling legislation to plan for exclusive guideways—with the provision that they be pragmatic, low cost, and incremental—the MTDB began detailed planning in 1976. The agency's search for a long-distance line that could be constructed at low cost and operated at high speed and low cost was aided by an act of God. A severe storm washed out portions of the San Diego & Arizona Eastern railroad, which follows much of the preferred alignment to the south. Mayor Wilson seized the opportunity in 1978, when the railroad petitioned the ICC to go out of business. The MTDB was able to purchase the 108-mile railroad for only $18 million. The agency decided not to seek federal funding, in part because of a

Transportation Finance in California

Users are the major source for funding transportation in California, as in most of the United States. California applies a variety of fees, primarily taxes on gasoline and diesel fuel, registration fees on cars and trucks, excise taxes on tires and other vehicle components, and fees on other vehicles. Nationwide, trucks pay a surprisingly large 31 percent of state highway user taxes and 46 percent of federal highway user fees. In California, the 1991 shares were estimated at 34 percent and 40 percent, respectively.[1] Fee/tax financing has many desirable qualities. It is generally simple to understand, the collection mechanisms are efficient, and it charges those who benefit the most. (Transit is a little more complicated.)

The one big liability is a political one. Legislators who support gas tax rate hikes to make highway improvements can be denounced for raising taxes.

California's 1959 freeway plan calling for more than 12,000 miles of free-ways—nearly one-third the mileage of the entire U.S. interstate system—illustrates the state's enthusiasm for freeways. However, freeway construction in California began a precipitous decline in 1967, primarily for financial reasons. (The environmental rationale came in 1975 during the Jerry Brown administration.)

The causes of the financial difficulties are familiar throughout the United States. On the cost side: inflation, growth in the scale of projects, rising land values, and the added costs of meeting environmental and community concerns. On the revenue side: a gas tax that is not indexed for inflation, improvements in fuel efficiency that allow people to drive farther on less gas, and the fact that California gets back less in federal payments than the federal government collects from the state. A financial package that had appeared generous in the 1950s was no longer adequate.

In 1971, California dedicated one-quarter cent of its 6 percent general sales tax to public transportation, creating the largest state transit subsidy in the country at the time. However, the governor and legislature failed to deal with the problem of financing free-ways, and thus effectively killed the 1959 freeway plan.[2] Localities were left to contend with severe transportation needs that traditionally had been handled by the state. Not until 1990 did the legislature submit a proposition to the voters to increase the gas tax by $0.09 a gallon. It passed.

Notes

1. Motor Vehicle Manufacturers Association of the United States, *Facts and Figures '93* (Detroit: Motor Vehicle Manufacturers Association, 1993), p. 81.
2. Brian D. Taylor, "Why California Stopped Building Freeways," *Access* (University of California Transportation Center, Berkeley), fall 1993, pp. 29–33.

---

recent federal decision that had denied funding to a rail project in Denver.

Completed on time and under budget, the original south line of the first new U.S. light-rail line in many decades was constructed at a bargain basement price of $7.3 million per mile. Ridership reached 30,000 per day in the mid-1980s, a level that initially had been projected for 1995.[18] Fares covered as much as 95 percent of the operating expenses until a second, 16-mile line opened. By 1995, fares were covering 66 percent of operating expenses. The trolley's capital costs translate into an investment per weekday roundtrip rider of less than $8,000. (By contrast, the blue line in Los Angeles was built a decade later for $50,000 per rider, a figure that does not factor in inflation.)

The MTDB's design philosophy—keep it simple and functional—has kept down costs. The agency has developed a reputation as a lean and mean construction manager that pays close attention to the bottom line. The trolley corridor is in an especially strong transit market. Tourists and travelers between San Diego and Tijuana make up 14 percent of the riders, and military personnel are frequent riders. The trolley captures the highest share of by-choice riders among all the San Diego transit systems: over 40 percent of its riders have an automobile that they could use for the trip.[19]

Despite the MTDB's good record in operating transit service, transit use in San Diego faces impediments. Some of these reflect prevailing development patterns. In 1993, the MTDB produced a manual, *Designing for Transit*, and joined the regional dialogue to reshape development patterns to better support transit.

Addressing Long-Range Transportation Needs. The integration of transit services, the construction of the trolley, and the shifting of growth to developed areas all helped San Diego deal with transportation and growth. However, the need for road investment and financing mechanisms was still large. The impacts of growth were on the minds of everyone, with the most prominent issue being traffic congestion. Although they complained a lot about traffic, residents were still surprised by a study from the Texas Transportation Institute that showed the growth of congestion in San Diego to be the highest among 60 regions studied. The San Diego region was hardly alone in its transportation funding crisis, as a 1984 survey by the County Supervisors Association of California of county road needs and financial deficiencies throughout California made clear.

The 1984 regional transportation plan was developed by SANDAG, Caltrans (the state transportation agency), MTDB, and local agencies. It identified projects and financial resources for the next 20 years. The price tag came to $9.9 billion, about equally divided three ways among local streets and roads, state highways, and transit. Not surprisingly, anticipated resources were estimated to account for only 60 percent of the needs. If San Diegans were to be spared a future of gridlock, it was essential to find new money to carry out the plan.

The primary source of funding for local and state roads in San Diego is the state gasoline tax (see "Transportation Finance in California" on the preceding page). Of revenues spent on state highways in 1983, 88 percent came from state and federal user fees.[20] Most of the expenses on city and county streets are for maintenance, and state and federal gas taxes (in 1983) paid only 40 percent of the costs.[21] In the 1980s, California had one of the lowest gasoline tax rates of any state and not enough revenue to sustain the state highway program.

To obtain enough money from a gas tax to carry out the San Diego region's transportation plan would have required a tax hike of $0.20 per gallon. A proposed $0.02 gas tax increase had been defeated in a 1982 election, and voters were not likely to support this much larger increase. Moreover, the transportation plan called for both highway and transit money, and Californians, like others, resist the allocation of road-user fees to pay for anything but roads. Less than 1 percent of California's highway tax revenue goes to subsidize public transportation. Furthermore, fuel taxes in California are collected at the refinery, not the gas pump, making a local gas tax assessment administratively difficult.

The transportation planners set their sights on a sales tax increase. A 0.5 percent increase in the sales tax would raise $2.25 billion over 20 years, enough to meet the funding shortfall, and it would be much more politically palatable.[22] Of course, a higher gas tax would have been a financial incentive to reduce driving, encouraging the use of transit. However, the primary goal was to raise money, not to shape demand. The SANDAG board sought state authority to create the San Diego County Regional Transportation Commission, with the SANDAG board serving as the commission. In 1985, the commission was authorized to determine a transportation sales tax of up to 1 percent, subject to the approval of a majority of voters.

In making the case for funding, it was important to communicate the message that the solution to growing congestion was transportation improvements, and not no-growth. The one-third each allocation to transit, state highways, and local streets and roads appeared to be roughly in line with the needs and, more importantly, opinion polls showed that the public approved. The one-third portion for city

streets and county roads was an attempt to deal with their severe disrepair, a consequence of Proposition 13, which had severely cut into local tax revenues. According to William Lorenz, an engineer for San Diego County: "Nobody got as much as they wanted," but the formula was about right. In fact, this allocation set a pattern for other California counties, according to Arthur Bauer, who heads Californians for Better Transportation.

Building support for the vote involved the creation of the private Foundation for Quality San Diego to help publicize the campaign through presentations to community groups and other organizations. "Prop A was designed to pass," according to campaign chairman James R. Mills, who had formerly served in the California senate. "We put together enough support for a majority. We left out projects with substantial opposition, which would cost votes."

The most popular part of the package, according to Mills, was the lowering of senior transit fares to 25 percent of the normal rate. (Senior passes then were available at half price.) Rail was generally more popular than highways, except in North County, where the spread-out population is more dependent on highways. Devoting 1 percent of revenues to bicycle projects was a popular idea.

Some needed highway projects were left out of the package because local pockets of opposition to them could have cost votes. One included project, the center section of SR-56, was opposed by local Sierra Club members. However, supporters of the transit package, arguing that opposition to the highway might hurt transit, persuaded the state organization to not take a position, leaving the local chapter unable to oppose the measure. Funding for transit operating subsidies, which continues to be a big issue, was left out because it lacked voter support.

Much of the financing of the $500,000 campaign to pass Proposition A was provided by developers and contractors. To have included a growth management element in the projects package would have lost their support, if not the support of the voters. The campaign was targeted through mailers. Voters in Chula Vista saw their local projects, and voters in Escondido saw theirs. The campaign also targeted seniors. The focus on mailings surprised skeptics. The campaign avoided TV ads not only because of the expense, but also because advocates did not want to offer a platform to no-growthers, some of whom were skilled at manipulating the media. The campaign obtained the support of virtually every elected official in the county. Official support was not publicized—because of citizen cynicism toward officials—but having it immobilized another potential source of opposition.

Proposition A passed in November of 1987, with a 54 percent majority. It allowed the imposition of the tax for a 20-year period. Some people were surprised that the vote ran best in minority communities not served by projects. This is, however, a common pattern in California (and probably elsewhere): the more conservative the area, the less the support for a tax initiative. James Mills feels that if one or two controversial projects had been included, 4 percent of the voters would have gone the other way—enough to defeat the tax. Unlike a similar campaign in neighboring Orange County that lost at the polls, the San Diego campaign "took polls and responded," according to Mills.[23]

Despite the fact that Proposition A dealt strictly with transportation improvements, and not growth controls, there was some related fallout. In Escondido, a North County community struggling to keep up with development, there were strong feelings that approving Proposition A could relieve traffic problems, but not if growth continued. The measure lost there. The next year, voters in Escondido adopted an interim development ordinance that capped residential permits at 430 units a year, an 84 percent reduction from the annual average for the two previous years.

Many counties in California have adopted sales taxes to fund transportation improvements. San Diego added its own twist by giving oversight responsibility to the existing regional association of governments, SANDAG, rather than creating a separate county transportation commission. SANDAG is also the region's designated metropolitan planning organization for federal transportation planning. SANDAG administers the funding through a regular meeting process. One-third of the revenue is allocated to projects in each of the three categories—state highways, public transit, and local streets and roads—that were specifically identified on the ballot.

A Network for Regional Cooperation on Transportation and Land Use. Concerning the relationship between transportation and land use in the San Diego region, there are many key players—18 cities, the county, two regional agencies, a variety of transit operators, the state transportation agency (Caltrans), and two local development agencies (the San Diego Unified Port District and the Centre City Development Corporation). A variety of formal and informal arrangements have helped foster cooperation and communications among them:[24]

■ **Cross-membership on policy-making boards.** The mayors and council members who decide on local planning and growth management issues sit on the boards of SANDAG, the Regional Transportation Commission, and the two regional transit

operators. Key issues often arise at the meetings of these regional agencies. Jack Limber of the MTDB points out that proposals for transit-related development projects are often floated at MTDB meetings. By the time they are presented to the city council, they are not a surprise and often some of the concerns have started to be addressed.

■ **Management groups.** Senior-level staff managers participate in a number of working groups that are sanctioned by policy boards. For example, a city of San Diego management group involving the departments of planning, traffic engineering, and development services meets regularly, as does a group of general managers of transit agencies.

■ **Shared staffing.** The MTDB has regularly used staff from the city of San Diego, providing valuable experience to the transit agency and facilitating coordination with the city. Several Caltrans staff members work at SANDAG.

■ **Task forces.** Land use and transportation agencies in San Diego often resort to task forces to deal with tough problems. An example is the Transportation Financing Advisory Task Force, which reported to SANDAG on recommendations for the sales tax.

■ **Contracting out to other agencies.** The level of trust among technical agencies in the region is such that they often are willing to accept the assistance of another organization with special expertise. SANDAG provides design services to Caltrans and travel forecasts to MTDB. MTDB has contracted with the city of San Diego for assistance with land use planning and traffic engineering along rail corridors. Local jurisdictions negotiate with developers to contribute to transit facilities, including rights-of-way for MTDB routes and, occasionally, stations.

■ **The telephone.** As reported in a 1989 survey, many transportation and planning agency officials often call someone from another agency who should be involved, and then handle problems within a team framework.

Outlook: Keeping on Track

San Diego dealt aggressively with its transportation needs, but not in time to avoid a voter backlash to growth. Between 1980 and 1987, weekday travel on surface streets grew by 30 percent and travel on freeways grew by 50 percent. But the number of freeway lane-miles increased only 8 percent. Traffic doubled on 20 different freeway sections. Part of this growth was due to an increase in the share of vehicles carrying solo drivers, from 81 percent in 1981 to 84 percent in 1984.[25]

The imposition of growth controls was soon followed by a deep recession. Much of the anti-growth ardor evaporated when the growth problem became more how to get it than how to keep it out. Of course, growth will eventually return. But planners have used the pause to develop a clearer vision of San Diego's future.

The same recession also has caused revenues from the transportation funding initiative to come up short. Planners have had to be creative in stretching the available dollars and lowering the expectations about which projects will be completed. The sunset of the funding in 2008 sets a deadline for addressing transportation needs.

Concerns about the economy have caused a perceptible shift in attitudes about regulating growth. Michael Stepner, urban design coordinator for the city of San Diego, says: "There is more interest in how to stimulate jobs, although people are still concerned about the nature of growth beyond the urban limit line." Allen Holden, the city's deputy director for transportation planning, observes that among staff members involved in development permits and regulations, there has been an attitude adjustment "from being antagonistic to trying to find solutions so as not to lose jobs. The city wishes to attract, not fight development, and has established a one-stop permit operation." Consultants have also noticed a difference. Arnold Torma, a transportation engineer with the firm of Katz, Okitzu & Associates in San Diego, notes that the city council is going out of its way to help projects: "We're scratching our heads, wondering, is this the same old council?" Holden observes that the timing of impact fees is a problem. If fees are to be used to finance growth, he says, they must be established early in the gowth cycle. Otherwise, substantially higher fees will have to be spread over fewer developments in order to raise the same amount of money needed for improvements.[26] San Diego's impact fees have recently been reduced.

Seeking a New Form of Development

Looking for ways to reduce the reliance on driving, the city of San Diego brought in urban designer Peter Calthorpe to help it formulate standards for development patterns. The city council adopted transit-oriented development and design guidelines—TOD guidelines—in 1992. To implement the guidelines, the city has focused on updating the zoning code, revising the street design manual, amending the general plan, and adopting a livable neighborhoods initiative.

Aware that conventional single land use, auto-dependent development patterns cause at least some of the traffic problems (as well as other problems), planners in San Diego are trying to "break the mold" of traditional development. Developers, designers,

and market researchers get locked into standard designs that have worked, and exacerbate the problems. According to James Mills: "Traffic is the biggest single factor in citizens anti-growth initiatives. In many parts of the county, we are building walled communities where citizens cannot walk to public transportation or anything else. We have allowed development without consideration of transportation options. We are no better than L.A.," complains Mills, in a particularly damning criticism for a city in which one anti-growth organization was called Prevent Los Angelization Now (PLAN).

Breaking the mold is difficult. One example is Horton Plaza, a project that reestablished major retailing downtown with good pedestrian and street connections. A traditional approach would have been to clear much of the surrounding land for parking, as in a suburban mall.

San Diego's Uptown District is another example. The project turned the site of a former Sears store into an urban and urbane neighborhood with a major food store, high-density residential development, and supporting shops and restaurants. The city had acquired the site in 1986 for a central library facility. When the city's plans changed, the community's residents and business leaders assumed an active role in determining the use of the site. The community wanted a supermarket and housing, and it did not want a minimall or a shopping center that would compete with the small established retail enclave in the neighborhood and cause more traffic.

The community and the city developed a plan, and the city selected a development team, a joint venture of Ted Odmark and John Thelan (residential component) with Oliver McMillan (commercial component). An extension of the existing street grid and a variety of architectural styles reflecting the eclectic nature of the neighborhood integrated the project into its surroundings.

The Ralph's supermarket that anchors the project certainly breaks the mold. It is one-quarter smaller than the company's usual store, it sits behind smaller stores, and most of its parking is underground (with a conveyor ramp for carts). Even so, the Uptown District Ralph's has ranked consistently among the top three performers in the chain. Ralph's has since built another store with underground parking and has a third under construction, while Vons, a competitor, is also building a store with underground parking in Hillcrest.

While working with the community took time and resulted in a more expensive project, it illustrates, according to Michael Stepner, that "an independent approach will result in a better product than one that is owner driven. A developer-driven approach

Detailing the TOD Vision

Rio Vista West, a $200 million, 90-acre mixed-use project in Mission Valley adjacent to the future trolley line is the first project to fall under the city of San Diego's new TOD guidelines. The developer, CalMat Properties Co., is a subsidiary of CalMat Co., one of the nation's largest producers of construction materials, which has used the property since the 1930s as a sand and gravel quarry. CalMat developed the 50-acre Rio Vista East project, which includes a Marriott hotel and approximately 400,000 square feet of office space.

Alan Nevin, managing director of ConAm Economic Research, points out that Rio Vista West is the first of only five or six large-scale, mixed-use projects that will be in the pipeline. "Unless you have the land at zero base and have very deep pockets, you really can't develop a project like this," he says. The plan calls for more than 1,000 units of medium-density housing, 325,000 square feet of retail space, and 25 acres of open space. Don Cerone, vice president of CalMat Properties, had participated in the development of the TOD guidelines and supports the general concept. Nonetheless, some points of contention have arisen:

- The developer became caught in a battle between city planners, who wanted narrow, walkable streets to support the design's neotraditional concept, and the public works staff,

who were concerned about substandard streets and potential liability problems.
- Citing the importance of clustering people around transit, the city and the Metropolitan Transit Development Board (MTDB) argued for a higher density than the developer had proposed. The developer was successful in arguing that 25 units per acre is the highest economically feasible density for this development.
- Although including a highway-oriented shopping center with a 120,000-square-foot discount superstore, a supermarket, and a drugstore seemed to go against the philosophical grain of a TOD, the center was critical to the economic feasibility of the project. In the interest of getting a TOD underway, the city was willing to go along, a decision that was made easier by the developer's incorporation of several design strategies to weave the shopping center into the fabric of the community.

CalMat was also successful in arguing that it should not have to subsidize rail construction costs like other property owners along the future Mission Valley trolley line, because it was providing right-of-way land to MTDB as well as a much more extensive street grid than a suburban style development would require. The streets will cost $2.5 million, which is more than the value of the land that was

contributed for the trolley line. Cerone points out that "the project needs to stand on its own, with or without the trolley." Not only is the effectiveness of transit in such a suburban location unproven, but also the transit construction schedule is very uncertain.

At a ULI workshop in San Diego in 1994, several area developers expressed concerns about TOD that tend to mirror Don Cerone's experience at Rio Vista West.

Access to transit, said many, may not be necessarily beneficial to a commercial project. For retailers, transit may provide access for a small share of shoppers but a large number of adolescents who may just hang out. The developer of one approved Mission Valley project is providing a right-of-way for the trolley, but it thinks that the new route might be an actual negative for the project.

TOD's reduction in parking requirements, say developers, may be a real benefit to residents who own fewer cars, but it can be a liability for markets with high levels of car ownership.

Higher density is a central tenet of TOD. Overcoming citizen resistance to density is a major issue. Bruce Sloan of EastLake Development points out that homebuyers are the ultimate public policy makers and he suggests that concepts like TOD should be marketed on the basis of project quality, rather than project density.

---

would have given a different result." Ted Odmark agrees that "too many projects are built just to be plain vanilla, because that's what everybody builds and that's safe. With this intensity of use, the project would have fallen flat if the design and execution were not outstanding."[27] Paul Buss, executive vice president of Oliver McMillan, says: "We were all surprised that 20 percent of Ralph's business is pedestrian, pretty unusual for Southern California." The condominiums seem to be pedestrian-oriented as well. Only four daily vehicle trips are made per unit. It is more typical for a project with 50 units per acre to generate an average six trips per day per unit.

The "transit" in transit-oriented development (TOD) does not have to be a train station. In San Diego, many neighborhoods proposed for TOD projects (TODs) will never have trolley lines. Bus streets will work too, and a major goal of TODs is to encourage

more pedestrian trips as well. Acknowledging the semantic problem, Stepner says: "We bought a brand name, but it has a much broader definition." TOD is a term that has been well established by Peter Calthorpe. Other descriptors have been tried. A local newspaper's "convenient, homey neighborhoods" is too wordy and cute. "Urban village" can apply to transit-related as well as pedestrian-oriented projects without a transit connection. TND (traditional neighborhood design) and NOD (neighborhood-oriented development) are other possibilities.

Will It Reduce Traffic? There are many hurdles to be overcome in implementing a transit-oriented or pedestrian-oriented development model. But it is not impossible to do so. Planners may view traffic engineers as inflexible, but in San Diego they have proved willing to recognize the benefits of a connected grid system and to rethink standard design practices. The real question is whether giving people a choice will actually reduce driving. Will people in TODs now walk to the store for a latte but then get into their cars to drive to work?

Planners have two choices for estimating how much a new development pattern can reduce traffic: look back to the future by examining current traffic patterns in developments generally built before World War II; or predict the future based on current mathematical traffic models adjusted to reflect higher densities and greater mode choices. San Diego has done both.

One study, by SANDAG, based on household travel logs collected in 1986, compared travel patterns of people living in 13 pedestrian-oriented or "traditional" communities with travel patterns in the rest of the region. The study communities had grid street patterns, a mixture of residential and commercial uses, and a pedestrian character—although none attained the level of pedestrian amenities that the city's new ordinance envisions for TODs. Six were located in urban portions of San Diego, and the rest were located in the urban portions of other cities. Because these communities were relatively old, the households were smaller than average and older than average—and both these factors are associated with fewer trips. Key findings were as follows:

- Households in traditional neighborhoods made fewer trips by each mode of travel—except bicycle—than other households. The difference was even greater for lower- and middle-income households, who made 11 percent and 13 percent fewer total trips, respectively. Higher-income residents of traditional neighborhoods came the closest in number of trips to other households, with only 2.5 percent fewer total daily trips.
- Households in traditional neighborhoods made fewer trips in cars, as well as fewer transit trips and walking trips. Households in traditional neighborhoods made about one-quarter fewer auto trips and transit trips, which were the modes where the largest differences were observed. Households in traditional neighborhoods made about the same number of bicycle trips as did other households.
- The second most frequent mode of travel after the auto was walking, a perhaps unsurprising finding considering San Diego's pleasant climate, but further evidence of the semantic problem of calling neighborhoods TODs when walking is seven times more likely a trip mode than the bus or train. In fact, the third most frequent travel mode was the bicycle, and it was used twice as frequently as transit.

Another study compared the amount of driving by residents of different San Diego neighborhoods, and showed large differences between the driving habits of in-town residents and suburban residents. While part of the difference may be a matter of demographics, as in the SANDAG study, the more suburban their place of residence, the more people drive. Residents of the Uptown District for example, logged about 16,000 miles annually on their cars, compared with residents of suburban Escondido, who logged almost 22,000.[28]

Will Anyone Believe It? The planning for Otay Ranch, the largest master-planned community in the region, shows that TOD planners may not believe what they preach—namely, that people will not need so many roads if they are provided with a range of good alternatives to driving. Plans for the 23,000-acre development southeast of San Diego were approved in October 1993 by the San Diego County Board of Supervisors and the suburban city of Chula Vista, the culmination of a five-year land use, facility, and environmental planning effort.

Otay Ranch follows a neotraditional plan, with 12 villages, each about one square mile, defined by open space and major arterials. Centrally located village cores will contain essential facilities and services: schools, civic facilities, day-care facilities, parks, and high-density housing. Six of these urban villages are located at future light-rail stops. A majority of the population of these six villages, which contain 15,000 of the development's total 21,000 housing units, will live within walking distance of the rail stops. A growth management program assures the provision of major facilities concurrently with development. Finding a density sweet spot was a challenge for the developer. TND-oriented planners argued that the proposed densities should be increased, while PUD-oriented planners thought they were too high.[29]

The analysis of transportation needs at Otay Ranch was a true Catch-22 for the developer. The California Environmental Quality Act offers no concessions for neotraditional planning in estimating the amount of traffic that will be generated by a development. The concept of the Otay Ranch master plan is to create a mix of uses and otherwise reduce reliance on the automobile, but the traffic analysis was based on traditional assumptions. The development may be transit-oriented, but Otay Ranch is still required to plan and pay for a full complement of roads.

Traffic calculations hypothesize a 10 percent to 20 percent reduction in traditional loads on adjacent arterials, but the developer is required to provide a standard six-lane right-of-way, although only four lanes might be required initially, according to Kim

Kilkenney of the Baldwin Company, the current developer.

Funding the transit system presented another problem. The MTDB approved a conceptual transit network for the South County in 1993, but as a new route it is not eligible for Proposition A funding. As part of an agreement with the county and the city of Chula Vista, the developer agreed to contribute right-of-way for the line and stations and a fair share contribution of construction costs. Construction costs have not been required of any other property along the corridor. Baldwin successfully argued that it should not be singled out. The transit agency is looking for additional contributions to assist in construction, possibly through broad-based benefit assessment districts. The developer has estimated that its land contribution amounts to 26 percent of the cost of the line, which is well above MTDB's rule of thumb that 20 percent of costs need to be borne locally.[30]

The Trolley Goes Uptown and Upscale

The MTDB's successful construction, operation, and gradual expansion of the San Diego trolley encouraged the agency to participate more broadly in joint ventures with the private sector and to assume a leadership role in regional development decisions that support public transportation. In its focus on the development and operation of transit services, the MTDB has demonstrated an entrepreneurial bent since the beginning, with agreements for food kiosks, childcare facilities, and parking lots on agency land. Building on the success of these projects, the MTDB decided to experiment with joint development on a larger scale.

To fulfill a need for office space for itself, and to offer a model for future private sector participation in mixed-use facilities at transit stations, the agency decided to develop an office building on a parcel that it had purchased for future use as a transfer station at the south edge of downtown. The resulting project, the MTS/Mills Building, was built in cooperation with the county of San Diego and a private developer. It opened in 1989 and has been very successful and widely acclaimed. Instructed to design an edifice that did not look like just another government building, the architects came up with a distinctive ten-story building that is set off by a 233-foot clock tower. An additional requirement, also met, was that the project pencil out.

The MTS/Mills Building was followed by the development (by the same developer) of the 34-story One America Plaza, San Diego's tallest building, on the site of a second transfer station on the north side of downtown.[31] This hands-on participation in development projects has given the agency credibility in its efforts to promote development patterns that are supportive of transit. Development proposals are sometimes presented informally to the MTDB before they are formally submitted for city review.

The trolley system covered 43 miles by 1995, including the 16-mile south line and the 26-mile east line, with its two-mile extension along the waterfront and the downtown leg of an extension to the north. Commuter-rail service—the Coaster—was introduced to the northern beachfront communities in 1995. Combined transit ridership (bus and rail) began to contract in 1992, possibly because of the recession. Farebox recovery—the percent of operating costs paid by riders—grew from 40 percent in the 1980s to over 50 percent, well above the U.S. average. A high rate of farebox recovery requires regular fare increases, which can lower ridership.

It appears that the affordable segments of the light-rail system are completed. The south line and initial segment of the east line, which were built largely along existing rail lines, cost under $8 million a mile, and the rest of the route to El Cajon cost just under $10 million a mile in 1989. For a ridership of 50,000, this translates into an investment of $3,000 per daily trip or $6,000 per roundtrip rider, one of the best bargains in U.S. light rail.

The next three segments—the continuation of the east line to Santee and the two downtown extensions—will cost substantially more: $28,000 to $32,000 per roundtrip rider. The planned new Mid-Coast line serving the high-income communities of Pacific Beach, Clairemont, Linda Vista, University of California at San Diego, and University City (one of the region's largest suburban office and retail centers) and a new Mission Valley line serving San Diego State as well as major attractions such as Old Town and the Jack Murphy Stadium will cost almost $30 million and $50 million per mile, respectively, and carry fewer riders per mile than the earlier lines. The east line extension is expected to cost $230 million and carrry 16,000 daily riders. This represents an investment per roundtrip rider of $29,000—cheap by U.S. standards, but considerably above what San Diego has paid in the past.

While these high-price transit investments are expected to generate large economic impacts, they are likely to undergo undergo considerable scrutiny because of the smaller levels of ridership, especially with growing needs for other public services in San Diego.

The transit market in these new corridors is quite different, as well. Downtown is the current destination for 75 percent of the region's transit trips, and the Mexican border is the destination for 27 percent (including some double counting of trips between downtown and the border). The new transit corridors are rich in income and jobs, and relatively far from downtown. People living in these corridors are not only less likely to use transit in general, but also less likely to work downtown, where transit is most competitive. To succeed in these new corridors, transit must attract by-choice riders, rather than riders who are transit-dependent. Current users are typically transit-dependent.

The Latest on the Transportation Plan

The new sales tax began generating transportation funding in April 1988, and was anticipated to collect $3.4 billion over a 20-year period. A plan of finance for the transportation sales tax program (TransNet) was established in February 1989. It is used to establish the approved schedules for implementation of the major projects, as well as the basis for the debt financing program. The majority of the local street and road programs are to be financed on a pay-as-you-go basis, while debt financing is needed to get the major projects underway quickly. The initial analysis determined that all of the major projects could be implemented over the program's 20 years.[32]

Eight bond issues between July 1988 and May 1995 raised $1.3 billion. A tax exempt commercial paper program was established in 1991 for such shorter-term financing needs as purchasing light-rail vehicles.[33] This program is like a line of credit, allowing SANDAG to pay off the notes when money is available at favorable rates. A refinancing program implemented in 1994 has the potential to produce savings of $15 million.

The latest update on the sales tax revenue picture projects that total revenues over the 20-year period will be $1.5 billion less than had been anticipated. To make up for the loss, other sources are being tapped. A new state/local partnership program provides matching funds at a rate declining from 18 percent in the first year (FY 1994) to 10 percent in 2001, when the program is assumed to end. Four new transit projects will seek federal funds, in a shift from the early philosophy of financing the trolley lines locally. Federal congestion management and air quality program funds will be used for transit projects. And half of the revenues from future increases in the state gas tax, assumed at $0.01 per gallon per year, will be dedicated for the transportation plan.

In 1991, project costs for highways were revised upward significantly, based on an independent review of the initial cost data from Caltrans and some changes in scope. Another update in 1991 revised the estimates downward by 3 percent because of the increasingly competitive construction market. Cost updates for trolley projects amounted to a 6 percent increase overall, although the cost of some projects was revised downward. The largest increase came in the western portion of the Mission Valley line, where increased mitigation costs, the loss of expected right-of-way from a cancelled development project, and the redesign of Jack Murphy Stadium as an aerial rather than a surface structure caused a 41 percent escalation. Such escalations concern developers along the right-of-way who have dedicated land for route alignments. They worry that the agency will not be able to deliver. In addition, time is a factor. This line was promised for the 1998 Super Bowl. One MTDB director, San Diego city councilman Ron Roberts, reports, however, that "the Super Bowl isn't going to rush us into decisions that would ignore environmental issues." The route's mitigation problems show that transit projects, however environmentally beneficial they may be regionally, must deal with the same types of local environmental concerns as highways. Roberts points out that MTDB is "caught between trying to correct one environmental problem without trying to create a second one."[34]

The transportation plan's financial analysis completed in 1994 made more conservative assumptions about revenues. (It was also the first regional plan in which costs were constrained to reflect new federal guidelines.) The analysis discovered that several important highway projects appeared to be in jeopardy. Several transit projects were delayed and the Mid-Coast line was deferred. Twenty percent of transit funds is spent on operating costs. This could be reallocated if other funding were found. Meeting the anticipated cash flow needs of local street and road programs, which rely largely on pay-as-you-go financing, poses no problems.

San Diego is already a leader in managing growth and improving transportation. A proposed HOV buy-in carries on this tradition. The buy-in would allow solo drivers to use an eight-mile stretch of I-15 in north San Diego County that is presently limited to high-occupancy vehicle (HOV) use during peak hours.[35] The buy-in would represent the first use of congestion pricing principles in San Diego—not counting the toll bridge to Coronado.

Congestion pricing is widely favored by economists and policy wonks as the most rational solution to traffic congestion. The concept has been less popular among public officials. In 1993, San Francisco was selected by the Federal Highway Administration to demonstrate the potential of congestion pricing. The project, charging higher tolls on the San Francisco-Oakland Bay Bridge during peak periods, seemed fairly innocuous, but it languished when not a single state legislator could be found to sponsor the necessary legislation.

Congestion pricing gives something to commuters rather than taking something away, which is what happens when general purpose lanes are converted to HOV lanes. The challenge for San Diego's HOV buy-in will be to keep the level of use on the HOV lanes low enough to not impair traffic speeds.

Quiet before the Storm

The recession and concerns about jobs have eliminated growth management as a pressing concern and also have depressed traffic levels. At the same time, transportation improvements have begun to make traffic better. "It's not a problem" is what people are now saying about both growth and traffic.

The trend in growth policy since the recession has been to promote, attract, and nurture business. The San Diego city council adopted an economic development strategy in 1992. Substantial progress has been made on regulatory reform and business attraction and retention.[36] The suburban cities have followed suit. A review of the city of San Diego's growth management system finds fault with its up-and-down execution, with the buildup of planning staff during periods of rapid development and staff cutbacks when development and development revenues decline. "In down periods, the city put planning on hold and in up periods it was too busy fighting fires."[37]

San Diego and the suburban cities are certainly not going to scrap their growth management programs, however. In 1995, the city of San Diego adopted the 1991–1998 housing element to improve the quality of existing housing and the availability of sites. One of the goals—to minimize government constraints without compromising the quality of government review—sends a clear message.[38] Support runs deep for preserving the qualities that make San Diego so attractive. Twenty years after *Temporary Paradise* appeared, a new initiative from Citizens Coordinate for Century 3 (C3)—*Toward Permanent Paradise*—revisits the earlier study and hopes to reenergize elected officials and develop a contemporary implementation strategy for the still relevant strategies that are set forth in *Temporary Paradise*.[39]

What's next for transportation? Completing the adopted plan requires continued support, and possibly adjustments. Changes in state and federal funding must be addressed. Certain projects not included in Proposition A may need reinstating. Because the sales tax measure addressed only those transportation needs that could be supported politically and financially, it neglected a number of key road project links that are now needed. Public support for the reinstatement of some of these projects is growing. The San Diego Highway Development Association, a group of highway boosters that has been in existence since 1935, has identified 17 major projects deleted from plans between 1962 and 1994 (12 of them since 1988). By 1996, another 13 projects were contemplated for deletion, five because of community opposition and the rest because of funding. The Highway Development Association has lobbied the regional agencies to impress upon them the significant air quality and congestion impacts of deleting adopted roadways.[40]

The funding period for the sales tax is approaching the midpoint. It appears unlikely that all the programmed projects will be completed. A look ahead to the following 20-year period needs to address unfilled promises as well as newly emerging needs. The California Supreme Court has thrown a monkey wrench into the process for extending the tax by ruling that such local taxes must be approved by a two-thirds majority of the voters. Such a margin of victory for a tax is unlikely.

Furthermore, the fact that traffic congestion has dropped way down on the list of regional problems makes it difficult to get people interested in even discussing transportation needs—much less in raising money for them. SANDAG's public facilities financing committee, which is made up of elected officials, has discussed imposing a regional public facilities financing fee on top of the fees that individual cities currently assess on developers.

San Diego's accomplishments in dealing with the vexing issues of accommodating growth—a lot of growth—and improving transportation are quite impressive. Clearly, however, the battle is not over. In fact, one worrisome question remains: Has success spoiled San Diego? Will the traffic problem be ignored until it once again reaches crisis proportions?

Notes

1. James R. Reading, *History of San Diego Highway Development* (San Diego: San Diego Highway Development Association, April 1985), pp. 3, 5, 7.

2. Donald Appleyard and Kevin Lynch, *Temporary Paradise? A Look at the Special Landscape of the San Diego Region* (Cambridge, Massachusetts: Massachusetts Institute of Technology, Department of Planning and Urban Studies, 1974).

3. Metropolitan Transit Development Board, *Designing for Transit* (San Diego: Metropolitan Transit Development Board, 1993), pp. 2–3.

4. Reading, pp. 13, 21.

5. San Diego Association of Governments, *Growth Monitoring Report and Forecast Update Schedule: Series 7 Regional Growth Forecast Monitoring Report* (San Diego: San Diego Association of Governments, August 24, 1990), pp. 3–4.

6. *Economic Bulletin* (Greater San Diego Chamber of Commerce), May 1995.

7. 1995 population estimates and 1994 retail sales from *Economic Bulletin* (see note 6), July 1993, p. 5. Employment data for 1990 from San Diego Association of Governments, *Interim Series 8 Regional Growth Forecast Allocation,* May 26, 1995.

8. San Diego Association of Governments, *Interim Series 8 Regional Growth Forecast Allocation* (San Diego: San Diego Association of Governments, May 26, 1995). One technical glitch that developed in distributing growth beyond 2005 based on adopted plans was that the amount of land planned for residential development was entirely consumed. A short-term fix was developed, but this indicates a more serious policy issue for the future.

9. 1990 Census data reported in *Economic Bulletin,* March 1995, p. 5.

10. J. Thomas Black, "Rising Residential Land Prices," *Urban Land,* July 1996, pp. 25–26.

11. Roger M. Showley, "Coming of Age: San Diego's Development Markets," *Urban Land,* April 1994, p. 27.

12. Pamela M. Hamilton, "The Metamorphosis of Downtown San Diego," *Urban Land,* April 1994, p. 32.

13. Centre City Redevelopment Corporation, *Horton Plaza Redevelopment Project Annual Report: Fiscal Year 1994* (San Diego: Centre City Redevelopment Corporation, September 1994).

14. Centre City Redevelopment Corporation, *Summary of San Diego Downtown Redevelopment Programs* (San Diego: Centre City Redevelopment Corporation, June 1992).

15. Lynne Carrier, "From Growth Management to Arrested Development," *Urban Land,* April 1994, p. 24.

16. Douglas R. Porter, *Profiles in Growth Management* (Washington, D.C.: ULI–the Urban Land Institute, 1996).

17. Peter M. Detweiler, "Is Cooperation Enough: A Review of San Diego's Latest Growth Management Program," in *State and Regional Initiatives for Managing Development: Policy Issues and Practical Concerns,* ed. Douglas R. Porter, (Washington, D.C.: ULI–the Urban Land Institute, 1992), pp. 63–64.

18. San Diego Association of Governments, *Trends before the San Diego Trolley* (Washington, D.C.: U.S. Department of Transportation, Technology Sharing Program, July 1982), p. 31.

19. San Diego Association of Governments, *1990 San Diego Regional Transit Survey* (San Diego: San Diego Association of Governments, 1991), p. 38.

20. U.S. Department of Transportation, *Highway Statistics: 1983* (Washington, D.C.: U.S. Government Printing Office, 1983), p. 71.

21. Ibid., p. 92.

22. William C. Lorenz, "San Diego Votes Yes for Transportation," *ITE Journal,* March 1989, p. 17.

23. James R. Mills (Proposition A campaign chairman), interview by author, April 4, 1996.

24. Eva Lerner-Lam, "Hardwiring Coordination in Transportation and Land Use Decisionmaking: Making It Work in San Diego, California," in *ITE 1989 Compendium of Technical Papers* (Washington, D.C.: Institute of Transportation Engineers, 1989), pp. 92–96.

25. "Freeway Volume Increases: 1980–1987," *INFO* (San Diego Association of Governments), no. 4, July/August 1987.

26. Allen Holden, interview by author, March 1996.

27. Information on Uptown District from Janice Fillip, "Uptown District, San Diego: Looking at the Future of Mixed-Use Development in American Cities," *Urban Land,* June 1990, pp. 2–7, and from Michael Stepner, remarks at a workshop on transportation, land use, and air quality (ULI, San Diego, April 1994).

28. John Holtzclaw, "Using Residential Patterns and Transit to Decrease Automobile Dependency and Cost" (paper prepared for California Home Energy Ration Systems by the Natural Resources Defense Council, San Francisco, 1994).

29. Anthony J. Lettieri, "Otay Ranch," *Urban Land,* April 1994, pp. 15–18.

30. *Otay Ranch: SPA One Public Facilities Finance Plan,* February 1996, pp. 3.2–3.9.

31. Jack Limber, *Public/Private Joint Development Partnership: San Diego MTS/James R. Mills Building Project Success through Self-Fulfilling Prophecy* (San Diego: Metropolitan Development Transit Board, no date), pp. 6–11.

32. San Diego Association of Governments, *1994 Plan of Finance,* (San Diego: San Diego Association of Governments, April, 1994).

33. San Diego Association of Governments, *TransNet Update: Transportation Sales Tax Program Progress through July 1, 1995* (San Diego: San Diego Association of Governments).

34. Mark Arner, "Trolley Line Must Bridge Funding Gap," *San Diego Union Tribune,* September 12, 1994.

35. "Pricing Paradigms: San Diego, California," *Buy-Lines* (U.S. Department of Transportation, Federal Highway Administration, Office of the Associate Administrator for Policy), spring 1996.

36. Price Waterhouse, "San Diego County," in *ULI Market Profiles: 1996* (Washington, D.C.: ULI–the Urban Land Institute, 1996).

37. Porter (see note 16).

38. Ibid.

39. *Toward Permanent Paradise: A Forward Look at the Special Landscape of the San Diego Region* (San Diego: Citizens Coordinate for Century 3).

40. San Diego Highway Development Association, *White Paper on San Diego County Transportation System* (San Diego: San Diego Highway Development Association, August 1995).

Chapter 10

Transportation and Development in Houston

Beyond Edge City

A building designer who comes to Houston and does not rent a car is not part of the solution; he is part of the problem . . . he is not being morally pure. . . . Going to Houston and not renting a car is like going to Venice and not renting a boat. It is missing the point.

— Joel Garreau, *Edge City*

While most economies shuddered, Houston's gloried in the oil shortages and price runups that spread around the world beginning in the 1970s. Business was great, money was made, and Houston was the Sunbelt's golden buckle. But when growth far outstripped the ability of the public sector to build transportation facilities and other infrastructure, traffic congestion became severe. Many people think that mobility, like youth, once lost can never be recaptured. Houston proved them wrong. But success in the past is no guarantee for the future, and Houston's future involves new challenges, new limitations, and new political leadership.

In terms of development and transportation, Houston is a contrarian. Rather than being troubled by automobile dependency, the region accepts it and unrepentantly makes massive road improvements. While the city of Houston is the only large U.S. city with no zoning, residents are protected through covenants and deed restrictions from the worst forms of adjacent use excesses. As in most growing communities, suburbanization is widespread, but compact citylike suburban centers also flourish. Unswayed by other cities' move to light-rail lines to improve transit, Houstonians turned down such a proposal and opted instead for a busway system, which has fostered not only transit use, but also carpools.

Among the highlights of Houston's experience with and approaches to regional transportation and development are that, as a progrowth community, it puts few restrictions on development and that, as an auto-oriented, outward-moving metropolitan area,

it boasts some of the largest and densest suburban activity centers—edge cities—that have ever been assembled. The Houston experience provides an outstanding example of a significant turnaround in investments in transportation and of the active participation of the business community. With the help of one of the nation's largest systems of HOV lanes and bus-based transit, Houston has succeeded in cutting back on traffic congestion.

Before the 1970s: A Foundation for Growth

Early Development

That Houston was founded by real estate speculators seems fitting. The Allen brothers bought 6,600 acres of land near the headwaters of Buffalo Bayou in 1836, the same year that Sam Houston's Texas army won independence from Mexico at the Battle of San Jacinto. They proceeded to lay out a town. The regular streets were 80 feet wide and the main east/west thoroughfare (Texas Avenue) was 100 feet wide. Houston served as the capital of the Republic of Texas from 1837 to 1840, a nice boost for business. By 1870, the population of the city exceeded 9,000 and Harris County registered 17,375, the second largest county in the state. In the same year, the U.S. Congress designated Houston a port, another boost and quite a feat for an inland city. The survey for a proposed ship channel to the Gulf port of Galveston was begun.

Figure 10-1
Population of Houston, 1970-1995

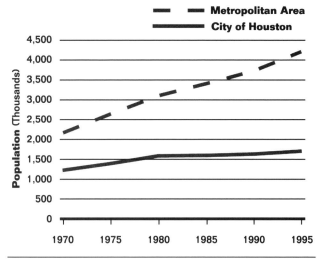

Sources: Decennial Census; and Houston-Galveston Area Council.

The turn of the century brought a new economic boost to Houston—oil. Discoveries at Spindletop and Humble put the city at the center of oil drilling and oil field equipment manufacturing. The opening of the Houston Ship Channel in 1914 facilitated the proliferation of refineries in the 1920s and 1930s. In the 1920s, the first building was air conditioned—a technological advance that was perhaps even more important for Houston's future. In 1929, the city planning commission recommended that a zoning ordinance be adopted, but it found no takers.

While the ship channel was by far the biggest transportation improvement, other forms of transportation also played a role. The first trolley cars, which were drawn by mules, appeared in 1868. In 1901, the Houston Left Hand Fishing Club brought the first automobile to the city; by 1905, there were 80. The development of a loop system for thoroughfares began with a 1941 master plan.

Postwar Development

After World War II, Houston turned its attention to making the civic improvements that a growing city needed and was an early beneficiary of the modern highway building era. Engineering for the Gulf freeway, Texas's first freeway, began in 1947. The management of growth seemed important to some people, and in 1947 the voters were given a chance to pass a zoning ordinance. They declined. The leading opponent was oilman Hugh Roy Cullen, who considered

zoning a socialist concept and denounced it as un-American. Growth continued throughout the 1950s. The region's population passed 1 million in 1955. In 1962, the move of NASA's Manned Spacecraft Center to Houston provided another big economic boost. The Astrodome and the Intercontinental Airport opened in the 1960s. In 1962, voters had another crack at a zoning ordinance, and they turned it down again.

Implementation of plans for a system of freeways progressed rapidly when the federal government established the interstate highway system in 1956 and Texas expanded its own highway building program at around the same time. New freeways combined with a modern arterial street system and the latest computer-controlled system of traffic signals offered Houstonians a very high level of mobility during the 1960s, and established a strong foundation for the growth that would take place in the 1970s.[1]

Current Conditions and Recent Initiatives: Zeroing In on Mobility

Houston Compared with Other Metropolitan Regions

Statistical indicators confirm what is intuitively obvious. Houston is a sprawling, low-density, auto-dependent region. Its urbanized area had a 1990 density of 1,800 persons per square mile, placing it, just behind Atlanta, as the second least densely populated large urbanized area. Even Phoenix ranked higher on the density scale. Houston drivers logged 26 daily per capita miles on the road, more than drivers in any region save Dallas and Atlanta. That the residents of Houston drive less and use transit more than the residents of Dallas suggests that the lack of zoning in Houston has not been responsible for the region's auto dependency.

Because Houston did not rank as high in terms of miles of freeway lanes per capita (seven regions were ahead of it in freeway lane-miles), it experienced significant and growing levels of congestion beginning in the mid-1970s. Moreover, Houston drivers were dependent on freeways for a high 58 percent of their travel on major roads; only five other regions were more dependent on freeways.

A regional congestion index developed by the Texas Transportation Institute indicates that Houston was one of only three major U.S. regions to cut down on regional roadway congestion in the 1980s, which was a period of major highway improvements for Houston. By 1991, Houston ranked 12th in congestion, just ahead of Portland, Oregon, and Detroit.

Figure 10-2

Transportation Indicators for Houston, 1970–1990

	1970	1980	1990	Percent Change	
				1970–1980	1980–1990
Population	2,169,128	3,101,293	3,711,043	43%	20%
Commuters	855,427	1,508,211	1,759,796	76	17
In Private Vehicles	741,565	1,381,989	1,594,796	86	15
Driving Alone	633,824	1,188,768	1,453,911	88	22
Passengers	107,741	193,221	140,885	79	–27
By Transit	42,885	42,903	72,617	0	69
Commute Share					
Transit	5%	3%	4%	–43	45
Private Vehicles	87%	92%	91%	6	–1
Travel					
Daily Vehicle-Miles Traveled (Thousands)	n/a	49,728	68,493	n/a	38
Annual Transit Trips (Thousands)	n/a	41,757	82,973	n/a	99
Median Single-Family Lot Values	n/a	$12,000	$18,000	n/a	50

Sources: Decennial Census; *Journey-to-Work Trends in the United States and Its Major Metropolitan Areas, 1960–1990* (Federal Highway Administration, PL-94-012); and ULI residential land price surveys.

Despite a heavy orientation toward private vehicles, transit use in Houston is not insignificant. With 31 annual transit trips per capita, the Houston region ranked 22nd among 35 large regions in transit use in 1990, slightly higher than San Diego, whose transit system has received much greater acclaim. In addition, long-term transit trends are up. Between 1980 and 1990, the number of annual bus riders doubled, compared with a 38 percent increase in personal vehicle traffic. Among commuters, transit increased its market share from almost three percent to almost four percent. Houston, along with San Diego and Phoenix, was one of the few U.S. regions to register an increase in transit market share in the 1980s. Transit's share is small, but it is growing. From 1990 to 1994, annual bus ridership grew by 1 percent. Since 1994, it has declined because of fare increases and fewer downtown jobs.

Development Market Trends

Because its economy was so linked to the oil business, Houston has been subject to a boom-and-bust cycle. During the oil bust of the mid-1980s, Houston lost 222,000 jobs, an extraordinary 14 percent reduction. In the 1980s, however, the regional economy underwent some basic restructuring and the importance of the energy sector in the economic base was reduced from 83 percent of basic employment to 60 percent, still high but no longer the only game in town.

In addition, shifts have occurred within Houston's oil sector. The volatile upstream operations (including exploration and oil field development) have declined from 80 percent of energy employment to 70 percent, as downstream operations (such as refining) have become relatively more emphasized. The upstream energy sector lost 184,000 jobs in the 1980s, but gained few back in the generally strong economic recovery.[2]

Houston is home base to the U.S. energy industry, serving as its administrative and engineering center. As a result, 19 of the country's top engineering and design firms have major operations in Houston. NASA's Johnson Space Center provides an additional engine for applied science. After energy and applied science, Houston is a leading center for medicine. The 600-acre Texas Medical Center is the largest concentration of medical practice, research and education in the United States.

Population growth has also recovered from the wild cycle of the 1980s. Between 1980 and 1990, the region's population increased by 20 percent, while the number of workers grew by 17 percent. Growth in outlying areas was even stronger. Fort Bend County to the southwest, registered a population gain of 72 percent. The city of Houston, with few annexations over this period, grew by only 2 percent.

A defining physical feature of Houston is its towering suburban centers, perhaps best symbolized by the Transco tower at Galleria/Post Oak, the leading such center. It contains more office space than most downtowns. Timing was a key factor in the development of these centers. As employers sought to cope with a shortage of workers in some of the administrative and technical fields making up Houston's core businesses, suburban locations near housing gave them an edge. The suburban centers grew explosively along with the suburban population.

The large Houston primary metropolitan statistical area covers six counties. With 594 square miles, the city itself is huge. Extending five miles beyond its boundary is a zone of extraterritorial jurisdiction, so Houston, unlike other central cities, is not surrounded by small incorporated cities. Lack of zoning is a third feature besides its huge size and jurisdictional control of boundary areas that distinguishes Houston from other major cities.

The region's 18 activity centers including downtown may be mostly suburban, but they are located within the city of Houston or its extraterritorial jurisdiction, so that Houston, unlike most U.S. central cities, participates in the wealth of the region's suburban centers. In 1995, the Galleria/West Loop area contained 26 million square feet of office space. Other major suburban centers include the Greenway area in midtown (with 12 million square feet of office space), the energy corridor along Katy freeway (20 million square feet), and the North Belt area near Houston Intercontinental Airport (10 million square feet).

Despite the impressive growth of the region's suburban employment and housing markets, downtown Houston is still a major presence—which is not well understood outside of Houston or, for that matter, inside. The CBD is the region's largest office center, with 33 million square feet of space. It accounts for one-quarter of the regional total, and 40 percent of Class A space. Inner Houston—the area inside the loop freeway—is growing in population, contrary to trends in most U.S. regions. Transit works best in this type of built-up area where trips are shorter and potential routes are many. This single demographic trend probably does more to keep the lid on driving in Houston than all of the transit system alternatives.

Downtown Houston has long been known as a nine-to-five downtown. Efforts are being made to renew the CBD and extend its hours of use. Over $100 million in public improvements are planned for the next three years. Approval for a 1,200-room convention hotel is expected in 1996, for a 1999 opening. The possibility of building new downtown sports arenas is under consideration.

In the meantime, suburban counties and municipalities continue to offer aggressive tax abatements to businesses locating within their boundaries, and these concessions are accelerating the decentralization of jobs in the region.

After the recent massive public investment in transportation, state infrastructure funding is concentrating on other areas. Major improvements continue to be made to wastewater collection and treatment, water supply, and storm drainage systems.

Houston's residential market is stable, as net migration into the region is relatively strong. Construction activity is strong in the area's ten largest master-planned communities. The region's single-family market is noted for its high levels of affordability.

The retail market is very active. Between 1990 and 1995, 15 million square feet of retail space was developed. Big-box retailers and large discounters have been the focus of this activity, and, as is true in other metropolitan markets, they are developing in an attempt to increase their market share rather than as a response to increased demand. Another 4 million square feet of discount and big-box retail space is expected to be added in 1996. The shift in focus from malls to power centers and freestanding stores has resulted in an oversupply of stores. Some regional malls will not survive the market changes.

For the office market, the road to recovery from the boom and bust of the 1980s has been long. Suburban markets have responded better than the downtown market. Three-quarters of the region's office space is in suburban activity centers outside the CBD. Less than 1 million square feet of new office space is expected to be built in 1996 and 1997.

The Port of Houston has helped make national and international trade an important sector in the region's economy. Warehouse and manufacturing space are the strongest segments of the industrial market. Most new construction is build-to-suit projects for local firms that are expanding or consolidating. Albertson's, a company that is moving to Houston, is building a 750,000-square-foot distribution center in west Houston. Michelin Tire recently built a 650,000-square-foot facility on the east side.

The Regional Mobility Plan

Very fast growth in the 1970s and early 1980s far outstripped the ability of the public sector to keep up with transportation and other infrastructure needs.

The region's early freeway development, which was generally in corridors with parallel arterial frontage roads and took advantage of the latest in computer-controlled traffic signal systems, had offered Houstonians a very high level of mobility during the 1960s. Nothing lasts forever, however, and the phasing down of the highway program combined with strong growth in traffic began to erode the mo-

Figure 10-3

The Relationship between Vehicle-Miles of Travel on Freeways and Lane-Miles of Freeway, Harris County, Texas, 1950–1990

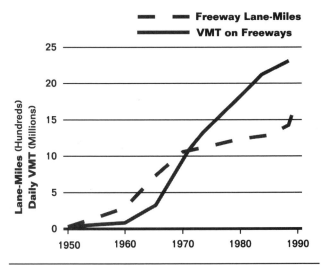

Sources: Regional Mobility Plan for the Houston Area: 1989; and Texas Transportation Institute.

bility that many had taken for granted. Figure 10-3 shows the different growth rates for freeway capacity and traffic. Between 1970 and 1980, freeway capacity rose by only by 22 percent, while daily miles traveled more than doubled. Average evening rush-hour freeway speeds dropped from 37 mph to 24 mph. Congestion spread throughout much of the area and much of the day. At one location on the Southwest freeway, so-called off-peak traffic filled the road during most of the time between rush hours. At the same time, an ailing city bus system was struggling for survival. In 1978, to prevent its demise, Harris County voters approved the creation of the Metropolitan Transit Authority (METRO) funded by a 1 percent sales tax.

Traffic congestion was Houstonians' leading worry. Houston's oil economy was attracting attention and the region was developing a reputation as a place with world-class traffic problems. By 1984, only Los Angeles had worse congestion. What to do?

Business to the Rescue. In 1981, the Houston Chamber of Commerce conceived a regional mobility plan. The planning effort was led by John Turner (head of Friendswood Development, Exxon's development arm). The goal was to "identify transportation projects which will provide substantially im-

proved mobility throughout the Houston area by reducing overall time and congestion," and a specific objective was to regain the mobility of the mid-1970s.

Private sector interests and all major transportation agencies were involved in the planning process, including the city of Houston, Harris County, the state Department of Highways and Public Transportation, METRO, and the Texas Turnpike Authority. A policy group (the Supergroup) and a technical committee of operating agencies that was chaired by Donald Williams, president of the Rice Center, worked together.

All kinds of transportation facilities and modes were considered by the planners, with the aim of achieving a unified, integrated, and coordinated transportation system for greater Houston. The interdependence of various transportation system elements—arterial streets, freeways and tollways, buses, high-occupancy vehicle (HOV) facilities, and high-capacity transitways—was explicitly acknowledged. Improvements were classified according to whether they were needed to meet today's demand, complete the system, or serve new growth.

The planners estimated the cost of achieving the mobility goals without limiting the budget to existing revenues. Unlike most transportation plans, this one was predicated on the belief that new sources of revenues would be required. The conclusion was shocking: to meet current transportation demands and accommodate 1 million new residents, Houston would need $16 billion over the next 15 years, and it expected to have only $9 billion. Sixty percent of the spending was needed to meet today's demands.

On the other hand, congestion was also costing Houston money. The annual cost of congestion to residents and businesses was estimated to be $1.9 billion. The 1981 Regional Mobility Plan identified options for raising $1 billion a year over the 15-year period to cover the $7 billion shortfall, including new gas taxes, an increased regional share of state funding, tolls, bonds, and the dedicated sales tax for transit improvements.

Acknowledging that no single mode, no single agency, and no government effort that does not involve business can succeed in improving mobility, the plan called for a full range of facilities. Highways would be considered the core of the transportation system, but even if cost were no object there were limits to highway construction. Some freeways could not be widened and some neighborhoods adjacent to arterials could not be disturbed. Thus, transit was needed to increase the passenger carrying capacity of the transportation system. The key elements of the Regional Mobility Plan (RMP) are as follows:

- **Additional roads.** Forty-two miles of existing freeways would be widened; 250 miles of new freeway lanes would be built; and 1,400 miles of new arterial lanes would be built. Twenty-six miles of toll roads are included because of funding realities.
- **Interconnections.** Houston relied on developers to construct portions of planned streets as they developed the adjacent land, but this method of constructing a street network often left a partially completed grid—and the gaps caused bottlenecks. The plan calls for "gap-filler" connections to improve the balance and continuity of the system, including the completion of 34 miles of freeways.
- **Bus services.** Several thousand buses would be added and associated maintenance facilities and park-and-ride lots built. The belief that Houstonians would never ride a bus had been shattered when METRO opened 14 park-and-ride lots serving downtown and other major employment centers between 1979 and 1982: most new services were operating at or near capacity within a month of opening.
- **Dedicated bus and HOV lanes.** Ninety miles of lanes on existing and new freeways would be dedicated to buses and HOVs. This would be the largest such system in the United States. At the time the mobility plan was being developed, Houston laid claim to being the vanpooling capital of the world, with over 2,200 vans in daily use. Inspired by successful HOV lanes in Northern Virginia and Los Angeles, the city of Houston and the Texas Highway Department opened a nine-mile contraflow HOV lane (reverse traffic flow on the inside lane of the road going in the off-peak direction) on the North freeway in 1979 that was successful beyond all expectations. It increased the number of people traveling in the peak direction by 50 percent in two years, which was the equivalent of adding almost two freeway lanes. With the success of this modest effort and the emergence of METRO as a well-financed transit agency, Houston was committed to the concept. It had become, according to the local press, HOV-positive.
- **High-capacity transit connectors.** Over 30 miles of facilities would be developed to serve the highest travel demand corridors and activity centers. These facilities would be connected to bus, HOV, and highway feeder routes. METRO was evaluating alternatives, and the mobility plan was left vague as to what mode of transit this would be. (The term "rail" was not used intentionally.)
- **Maintenance and replacement.** A budget allocation for maintenance and repair, items frequently ignored in transportation planning, was specified to ensure that the plan's new facilities would maintain their function.

According to Roger Hord, vice president of the Greater Houston Partnership, which evolved as a partnership between the business community and the city, one of the major accomplishments of the RMP was that it "focused official and public attention on the problem, and offered marketing to make the case for tax increases." The need for additional funding for the region's ambitious transportation programs became widely accepted, first within the business community and then in government circles and among the public. Also accepted, as an article of faith, was that the growth that was the cause of much of Houston's transportation problem would continue. The region's sharp economic downturn began in 1982, soon after the presentation of the RMP. Nonetheless, the impressive, coordinated effort to raise funds continued.

State funding is Houston's largest source of transportation financing. At the time the RMP was presented, the state motor fuel tax, the nation's lowest, was the largest source. The amount of money available from the state has grown considerably since the early 1980s—aided by the $0.05 a gallon increase in federal gasoline taxes in 1983 and a similar increase in state gas taxes in 1984; substantial increases in vehicle registration taxes in 1984; and a more equitable allocation of state taxes. State and federal funds now contribute between $350 million and $430 million annually. A kicker came in the form of an amendment to the 1991 federal transportation act by Texas Senator Lloyd Bentsen. The share of federal transportation taxes collected in Texas that are returned to Texas had been 75 percent; the amendment increased this share to 85 percent, for an extra $200 million—of which Houston receives the largest piece.

RMP advocates successfully resisted several attempts to curtail the transit sales tax passed in 1978. It currently generates $180 million annually. However, it pays for a range of highway services that are normally considered beyond the responsibility of a typical transit operator.

In 1984, city of Houston voters approved the largest public bond program ever undertaken in Houston, of which $295 million went to street, bridge, and traffic control programs. Street and bridge bond elections in Harris County raised $201 million in 1982 and $255 million in 1987.

Voter approval in 1983 of the Harris County Toll Road Authority came with an authorization for $900 million in bonds. The toll roads, says Harris County Judge Jon Lindsay, who led the campaign, have been "an absolutely essential spark and component of the mobility plan," stimulating the Department of Highways to build the related feeders and interchanges. The Toll Road Authority accelerated the development

Figure 10-4

Urban Traffic Congestion in Houston, 1975–1992

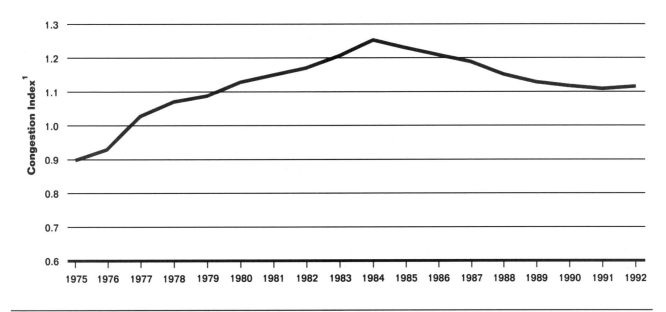

[1]TTI's roadway congestion index includes freeways and principal arteries and takes into account the effects of computerized transportation management systems. Levels above 1.0 indicate undesirable congestion.
Source: Texas Transportation Institute.

of key projects. Toll projects, according to Roger Hord, accounted for 45 percent of the new lane-miles added from 1982 to 1989, which helped Houston defend against complaints that it was getting an unfair share of state funds. That user fees paid for them served as an important defense, says Lindsay, against the belief that Houston was getting all the state's money, when in fact the region was asking for only the 30 percent of state transportation revenues that it had generated —with no retribution for years of neglect.

The RMP at Midterm. Cumulatively, total regional transportation spending for the seven year period between 1982 and 1989 was $6.5 billion, and annual expenditures during the last three years of the period exceeded the RMP's target of $1 billion per year. Through 1995, $11 billion had been spent— of the $16 billion that was estimated to be needed. The RMP steering group has rededicated itself to the achievement of the annual $1 billion funding goal through 2000, and plans to add $2 billion of private and toll projects.[3]

A report issued in 1989, midway through the planning horizon, said that the RMP was on track to meet its goal of recapturing the mobility that Houston enjoyed in the 1970s. Perceptions from outside the region were that a sinking economy had solved the traffic problems. However, there was never a year in which traffic actually declined, according to Tim

Lomax, a research engineer with the Texas Transportation Institute, a part of Texas A&M, who provided technical assistance to the study. Between 1982 and 1989, vehicle travel in the region actually increased by 10 percent. But road projects that had been completed or were under construction added 43 percent to freeway capacity and 27 percent to arterial capacity in this period. At the same time, the transit system expanded its service in miles by 57 percent, and increased ridership by 65 percent.

Two major toll roads had been built—the 21-mile Hardy Toll Road paralleling I-45 toward the Houston Intercontinental Airport, and the 28-mile Sam Houston tollway to the west near the planned outer beltway. Nearly 46.5 miles of a planned 95 miles of dedicated bus and carpool lanes (which Houston calls transitways) had been completed. And private developers had kicked in over 350 lane-miles of arterial streets, 30 percent of the total added. Kathy Whitmire, Houston's mayor, told ULI members that "elected officials never swayed from the commitment to improvements in infrastructure, which in addition to transportation improvements included water projects and a convention center."[4]

The results of these investments could be readily seen. Average freeway speeds during the evening peak period increased from 38.3 mph in 1980 to 49 mph in 1994—a 28 percent increase. The number of

The Nonrail Transit Alternative

The planned bus/HOV transitway system will eventually include special facilities for buses and carpools on six freeways into downtown Houston. Transitways potentially increase the person carrying capacity of freeways. A 1991 study showed, for example, that the system's four HOV lanes then in use carried from 39 percent to 129 percent more people during peak hours than did the general purpose lanes. Average bus operating speeds doubled—from 26 mph to 54 mph—which is a substantial benefit to riders and the operator. The savings in bus operating costs were estimated to be $4.8 million.

The four transitways carried 17,000 daily bus riders, which is close to the ridership on some new rail systems. Bus riders made up only one-quarter of the daily riders/passengers on the transitways. Originally designed for buses and vans, the HOV lanes were opened to carpools in 1986, a change that has been a huge success. For the bus riders only, an estimated $3 subsidy per trip was required to cover the operating costs. The operating cost per bus passenger-mile was estimated to be $0.13, which is more than on San Diego's light-rail system but less than on Portland's or San Jose's. The capital costs for the transitways up to 1991

were $6 million per mile, which is half the cost of the three light-rail systems just mentioned. The study's comparison of the transitways to light-rail systems was not intended to show that one kind of project was good and the other bad, but it must have made Houston's transit planners more comfortable with the choice they had made.[1]

Note

1. Dennis L. Christiansen and Daniel E. Morris, *An Evaluation of the Houston High Occupancy Vehicle Lane System* (Washington, D.C.: U.S. Department of Transportation, June 1991), pp. iv, 19–22.

---

miles of arterial streets that are severely congested was reduced from a 1985 peak of 74 percent to 29 percent in 1992. The area within a 30-minute drive of downtown almost tripled between 1982 and 1991, making Houston one of the few U.S. downtowns—and perhaps the only one—to have increased its primary trade area in recent years. The rush-hour trip to the airport from downtown that seven years before took almost 50 minutes, if there were no traffic incidents, was reduced to a consistent 30 minutes on the Hardy Toll Road.

By 1991, congestion levels were at their lowest since 1979, according to the TTI index, representing an improvement of more than 10 percent since 1982 (see Figure 10-4). This improvement was greater than that in any other of the 27 large urban areas tracked by TTI, among which only two other urban areas recorded any improvement at all.

Public transportation in Houston has come a long way. An underfunded city bus operation has changed into a well-financed, award winning regional transit authority. Transit plays an important, albeit relatively small, role in the regional transportation system. The number of bus riders doubled during the 1980s, which is one of the highest growth rates among major U.S. regions and stands in sharp contrast to national trends. On-time performance went from poor (only 39 percent of trips were on-time) in 1982 to excellent (97.8 percent) in 1988. METRO served 6 percent of all regional work trips and 1.6 percent of all trips in 1990.[5] A 1993 survey of large employers estimated that 38 percent of the workers in downtown commute by bus, which is a quite respectable share for a city where, outsiders believe, no one rides the bus.[6]

Houston's 95 miles of dedicated bus and car pool lanes will be the world's largest transitway system. At the end of 1990, a total of 46.5 miles of HOV lanes served 67,000 persons each day, a 50 percent increase over the previous year. This unusual hybrid project —part highway, part transit— resulted from "the synergy between transit and highway agencies working together," according to Tim Lomax. (See "The Nonrail Transit Alternative" on this page.)

The only component of the RMP on which little progress had been made at midterm was the rail project. In fact, as regards rail there was a significant midcourse correction in the RMP. As time went on, this one unfinished aspect of the transit—the high-capacity rail connection—had become a source of frustration, controversy, and political turmoil.

When the RMP was under development, it was expected that in a few high-density travel corridors rail service could be justified. Moreover, many Houstonians considered that having a rail system would be in keeping with Houston's emerging status as a world-class city. The first transit defeat came in 1983, when voters turned down a proposed 18-mile, fully grade-separated rail project. The METRO board regrouped and reassessed, and in 1986 adopted a regional transit plan that included a rail "system connector" linked with the HOV lanes to serve the major employment centers. An extensive campaign followed. In January 1988, voters approved a plan shifting 25 percent of the transit-dedicated sales tax receipts to a general mobility fund and authorizing METRO to implement the rail plan.

In the meantime, however, two national studies that were critical of recent rail projects came out. METRO's chairman, Bob Lanier, began to question the rail plan, and thus irked Mayor Kathy Whitmire, a rail supporter, who subsequently appointed another chair. In March 1991, METRO opted for a monorail project (as the "locally preferred alternative," which is transitspeak for a rail project seeking federal funding). By this time, Lanier had decided to run for mayor himself and he defeated the incumbent. Finally, in February 1992, with the help of the new mayor's appointees and after 17 public hearings, METRO withdrew the monorail proposal and substituted a regional bus plan. Warren Dold, a Friendswood Development executive who wrote the original RMP, called the elimination of the rail project "a midcourse correction of significant note."

The Chamber of Commerce, an early supporter of the rail project, continued to stand firm after much internal discussion. The inclusion of a "high-capacity transitway" in the original plan had been a vague reference to rail. "We couldn't get anyone from METRO to say there was a rail proposal," according to Roger Hord of the Greater Houston Partnership. Interestingly, the RMP was not mentioned in the 1983 campaign for rail that was defeated, although it has been invoked in every other major transportation decision in Houston except for restarting the space shuttle program. The most recent battle put the Chamber in a painful position: its carefully crafted coalition came unglued and it found itself on the opposite side of the issue from Lanier, the former chairman of its mobility committee. The Chamber took some lessons from this experience: vigilance is as important as the initial agreements. When things are going your way, you still cannot relax. And, relationships in politics will not necessarily stay the same. Hord decries "the years thrown away during the 1960s and 1970s, when growth was high and infrastructure improvements few, by arguing over rail plans," meaning that it would have been wiser to focus efforts on roadway improvements.

The Real Estate Benefit. If the RMP had been a normal transportation planning exercise, it would have included extensive studies of land use with little solid evidence of real estate benefits. Because of the pressing need for improvements in the short term, however, the planners did not delve into the details of future land use scenarios. (Houston's lack of zoning would have made this a pretty academic exercise anyway.)

But some excellent examples of the benefits of the transportation program to property owners and developers—in spite of the economic downturn—can be found. Greenspoint, a project of Friendswood Development, one of Houston's leading developers, is one of the better examples. Development of the 220-acre, multiuse project was started in 1978 in a northern location adjacent to I-45 and convenient to Intercontinental Airport. Access to the site from downtown and most northern residential areas was very difficult because of congestion on existing north/south routes and the absence of any east/west connections.

The RMP planning effort produced two roads that offer access to Friendswood's holdings on I-45 —the Hardy Toll Road running north/south to the airport and the Sam Houston tollway running east/west. Says Mark Preston, the Grubb & Ellis leasing agent for the project: Friendswood "had vision and could see where roads needed to flow. With the infrastructure here, Friendswood's ten- to 12-year-old investment is now paying off."

On the west side of town, the suburban markets adjacent to the Sam Houston tollway saw a marked increase in activity following the opening of the road's third phase. "There's no question the tollway has been a boon to that area," Charles Gordon, cochairman of Trione and Gordon, told *Commercial Property News*. Michael Gorney, president of Vantage Houston, reported that improved access has pushed Class A office rents up from $10 per square foot before the tollway opened to an average that exceeded $13 a square foot in spring 1992. The Sam Houston tollway made two office submarkets—I-10 and Westchase—that were separated by distance into a single market that can be traversed quickly.

Outlook: Beyond the Regional Mobility Plan

The 1989 review of the RMP by the Greater Houston Partnership's Committee on Regional Mobility documented a stunning record—one that is probably unmatched in any other large urban area, although unfortunately no one keeps such scores. And, it identified projects that were necessary to meet Houston's mobility needs through the remainder of the century.

According to the plan, an additional $5.7 billion above the money spent through 1989 will be needed to achieve acceptable levels of mobility by 2000.

Although the proposed rail connector project has been scrapped, the plan's commitment to transit is strong. Transit accounts for 21 percent of the transportation investments that have been made, and for 20 percent of planned investments. The Partnership's review estimates that $4 billion will be required for the maintenance and operation of transportation systems from 1989 to 2000. Research shows that proper maintenance strategies save money over the long run, money that can be invested in further improvements.

The 1989 review also made some suggestions for some new approaches to traffic issues for Houston.[7] One was strategic thoroughfares, which are arterial streets with design and signal improvements that can increase travel speeds. This concept, which is also under development in Chicago and Orange County, California, would provide relief for the freeway system, which carries 40 percent of travel in Harris County on 3 percent of the route mileage.

Another suggestion was for computerized transportation management to improve the efficiency of existing freeways, tollways, and adjacent arterial streets—by as much as 25 percent in congested sections. A related measure would be to incorporate traffic signals into a high-tech network of traffic signals throughout the region that can be operated together for more efficient travel.

The report also suggested that Houston develop a multimodal congestion index that reports on all travelers, not only on cars. All communities that try to use level of service (LOS) standards to measure congestion are limited by its exclusion of nonautomobile travel. The congestion index was very important in making the case for transportation improvements in Houston. Making it multimodal would support the region's commitment to a multimodal transportation network.

Concerns about traffic and transportation moved to the back burner when the last recession raised more basic concerns about the economy, the community, and quality of life. Funds intended for transportation were put to different uses in a pattern that may continue as the economy shifts. And the subject of zoning has been brought forward again.

What will the next round of transportation improvements look like? Will the region's decision makers continue to stay the course on completing improvements? Or, lulled into complacency by their successes, will they let their attention wander? If some recent developments are any clue, aggressive building programs may be replaced by more aggressive management efforts, and some rethinking of growth patterns may take place.

Broadening Transit's Role

Bob Lanier, the mayor who campaigned as an opponent of a rail system, saying that a bus system would serve Houston better, was able to divert 25 percent of METRO's (transit) money to a general mobility fund that pays for street repairs and traffic signal maintenance. The eminently logical basis for the diversion is that that such programs can improve bus travel. Of course, they serve motorists as well—a much larger constituency.

This type of funding flexibility was envisioned by the architects of the federal ISTEA legislation, although they envisioned the flow going the other way. Houston was able also to redirect some of its street repair money to law enforcement, an issue with higher visibility in the 1990s. Some transit supporters saw the diversion of METRO funds as a raid on the kitty that had been built up for the rail project. However, it seems to be a dead issue. The mayor's action was highly popular in some circles, and he has been reelected three times by wide margins. In fact, the diversion action may have saved the dedicated sales tax, which has been under occasional scrutiny from people who think it is unnecessarily high.

In 1994, the transit agency was officially given a broader role: to "meet the public's diverse needs for improved mobility by providing public transit, traffic management, and infrastructure improvements in partnership with other public and private entities." METRO claims to return value to the region through its transit services, transportation infrastructure, and "intelligent management of general traffic flow." METRO's assumption of a broader transportation role in an expanding market puts it in an unusual position among urban transit agencies, most of which have declining bases.

Working Together to Manage Traffic

Traffic management centers exist in other metropolitan regions, but Houston's TranStar Center is the first to bring highway and transit personnel under one roof. The area's transit agencies and highway agencies have already worked together on planning and building facilities, like the transitways system. Now they are embarked on a project to jointly manage facilities on a day-to-day basis.

The TranStar Center opened in 1996 as a joint venture of the city of Houston, METRO, Harris County, and the Texas Department of Transportation. TranStar is responsible for the planning, design, and maintenance of transportation operations and for the management of emergency (police and fire) operations within the greater Houston area, which extends over 5,436 square miles. Among the center's initial responsibilities are the management of traffic on 300 miles of freeways and the management of 2,800 signals on arterial streets.

The executive director reports to an executive committee consisting of the senior representatives of the participating agencies. This committee must work out any interagency disagreements to keep the center functioning smoothly. When these key transportation officials meet to resolve management issues, they also discuss and informally resolve longer-term planning issues. Gary K. Trietsch, Jr., a district engineer for the state DOT, is impressed with how well the agencies work together in this setting. Such cooperation and coordination, he says, "is very unusual in such a large area."

A Transportation Strategy for Uptown Houston

Uptown Houston, an upscale area built around the Houston Galleria, is said to be the nation's largest suburban activity center. With almost 24 million square feet of multitenant office space in 1989, it was larger than downtown Denver. It included 3.9 million square feet of retail space, 4,500 first-class hotel rooms, and 4,000 apartment units. Its daytime population was about 220,000.

Unlike most downtowns, the Uptown Houston business district is served by a limited roadway system that was originally intended to serve a primarily residential area with some supporting retail and office development—one of the legacies of Houston's lack of zoning. By the late 1990s, the roads had been expanded almost to the limits of the rights-of-way. Transit service was limited and oriented primarily to serve downtown, so fewer than 2 percent of trips into the district were made by transit.

Since the founding of the Uptown Houston Association in 1975, transportation issues have been at the top of its agenda. Major employers have helped develop vanpool programs. Many employees work flextime to avoid travel peaks and some work shifts start at 7 a.m. Developers have funded some transportation improvements on a voluntary basis. Uptown Houston was a key destination on METRO's proposed light-rail system.

In 1987, the state legislature created Harris County Improvement District #1, allowing property owners to tax themselves to make improvements. With the assistance of a planning grant from the federal Urban Mass Transportation Administration, a transportation improvements strategy was developed based on an understanding between private sector landowners and employers and the public agencies that access and circulation had to be considered for the area as a whole. The improvements strategy had five components: arterial streets, freeways, transit, pedestrian circulation, and transportation management. The estimated cost was $111 million.

At the beginning of the process, private sector participants tended to think that the public sector should solve the transportation problems and that the solutions were long overdue; and public sector representatives tended to think that the developers had created the problems and should stop developing projects that would exacerbate them. After much discussion these attitudes evolved, and both sides accepted responsibility for solving the problems.

METRO was considered to be a prime source of potential assistance. The agency had been spending one-quarter of its dedicated sales tax revenues on roadway improvements, but had made no improvements in the Uptown Houston area. In 1987, METRO estimated that the area needed $20 million for roadway improvements, and, in 1989, it approved a concept that allocated almost $45 million of the proposed $111 million program for arterial streets. Private sector interests were to pay 40 percent of the costs of arterial improvements, primarily through the dedication of rights-of-way. The remainder of the costs would be picked up by the state transportation department. At this time, METRO was still planning a light-rail line through the corridor, and private sector interests, which had helped plan the route and station locations, would assist in acquiring the rights-of-way.

The cancellation of the rail project and the subsequent recession have since put most of these transportation improvements on hold. However, some important principles have been established. One is that areawide planning is necessary, in the same way that downtowns need to take a comprehensive approach to planning. Another is that the private sector and the public sector both need to participate in financing the improvements. And third, the improvement district offers a useful vehicle for implementing a district transportation program.[1]

Note

1. See Uptown Houston, *Comprehensive Transportation Strategy: Final Report* (Houston: Uptown Houston Association, 1991), pp. 1–13.

Looking to 2010

In 1989, the region's metropolitan planning organization, the Houston-Galveston Area Council (H-GAC), adopted a long-range plan, known as Access 2010. The plan was revised in 1994 to reflect new federal rules.

Anticipating that the population will grow to 5 million (from 3.7 million in 1994), the plan called for significant expansions of the freeway system and arterial road system, expansion of the region's HOV network to 157 miles to accommodate transit growth, and a much greater emphasis on bicycling (for which it proposed 335 miles of new bikeways) and walking.

Most amazingly, the planners determined that the region had sufficient available resources—assuming a continuation of the flow of funds from the sources that were being tapped by the RMP—to pay the esti-mated $16.5 billion price tag.[8] H-GAC's financial plan, a federal requirement under ISTEA, suggested that no aggressive steps would be required to meet the region's transportation needs.

However, the future may not be that rosy. The costs of some new projects—including some expensive road widenings and reconstruction—have come in at significantly higher prices than had been expected. A planned widening of the Katy freeway serving the fast-growing area to the west, for example, is expected to cost $1 billion.

At the same time, some of the funding sources may not deliver the revenues that had been anticipated. Houston may not get its expected share of state and federal transportation funds. Federal appropriations have been constrained, and improved fuel efficiency means lower gas tax revenues from both federal and state sources. Increased competition from other areas has reduced Houston's share of state funds from a high of 26 percent when Lanier was METRO's chair, to 19 to 20 percent currently. In addition, there is competion from other state agencies. Texas now finances the state police out of gas tax revenues, and recently the state DOT acquired excess prison property to help bail out the prison system.

According to Alan Clark , the H-GAC's transportation director, Houston "may have turned the corner on expanding highways. Management and operations are now as important as building new capacity. The highway system is maturing, as we build out to eight to ten lanes in most locations." All highway solutions are expensive. In the Galleria area, for example, the H-GAC has determined that the loop freeway cannot be expanded because of adjacent commercial development. However, the frontage roads could be tunneled, a solution that will probably be implemented. The $150 million price tag is believed to be reasonable for improving access to such an important business district. Clark proposes less grand solutions: complete the arterial network, make intersection improvements, and manage traffic at a higher level. On this last approach, the TranStar Center is one step, and another is Houston's current installation of intelligent transportation infrastructure to make roads smarter, not longer.

Living in the Central City

The city of Houston has embarked on programs to improve the downtown and neighborhoods in the central area. Mayor Lanier, a former developer, recognizes the importance of having a close-in population base to support the downtown. "My experience is, you get people living somewhere first, and the shopping and entertainment will follow."[9] He has established a nuts-and-bolts program called "neighborhoods to standard" for stabilizing neighborhoods and competing with suburbs for middle-class families. The emphasis is on such details as street repairs, sidewalk improvements, and street lighting.

Representing the business community, the downtown association has been advocating public improvements for stabilizing the downtown for a number of years. Houston is focusing on the population base, rather than on the expensive facilities like sports arenas that other cities favor as redevelopment generators.

The program to strengthen the residential base seems to be succeeding. "We're gaining population inside I-610 at the rate of 5,000 people a year," says Lanier.[10] This result would please many cities with aggressive growth management programs to encourage infill development. The program has transportation fringe benefits. It helps promote transit and reduce driving. The central city is METRO's prime market area. Ninety percent of the transit trips made in this area are local trips, mostly around town; only 10 percent are for regional commuting trips, which generally involve suburbanites driving to park-and-ride lots to take a bus downtown. Close-in residents also tend to drive less because they are near downtown and other employment centers.

Maintaining Funding Support

Keeping the transportation money flowing is a concern. In Roger Hord's view, the system is now on autopilot. Agreements have been forged for the allocation of state and local funds, which will guarantee Houston's fair share. The transit system that has developed, says Hord, will drive the stake through the heart of any future rail proposals—not only will it be virtually impossible to take the transitways out of service to install rail lines, but also high-speed buses to downtown can compete against any rail line in terms of travel speed.

Despite this view, the possibility that competition from rural interests and other cities will erode state funding is a real one. It would not be surprising to see funding shift to other metropolitan regions and rural areas. In addition, because the transportation investments have paid off and things are getting better, transportation has ceased being the most important concern of the community. Crime is a growing concern (which was neatly linked to transportation when the diversion of METRO funding to street improvements freed up local street money for police programs). John Walsh, former president of Friendswood Development who became president of the Houston Partnership in 1995, says that the decision to put METRO's money into concrete, police hiring, and the maintenance of basic systems was the right one for the time—as the mayor's wild popularity (a 90 percent approval rating) shows. "It is now time to catch our breath, and decide what to do to position Houston as a world-class city in the 21st century. The two mayors, Whitmire and Lanier, have had such an impact on local policy that, with Lanier's term limit approaching, the business community will need to regroup." Walsh is concerned about the long-run impacts of the diversion of transportation resources. A thoughtful look at Houston's future is in order, says Walsh, in light of fundamental changes in the economy and regional demographics, and also because Houston has "spent tomorrow's money for today."

Lessons

When he became district engineer in 1995, Gary Trietsch, was struck by how effectively the region's transportation agencies and the Chamber of Commerce work together. If the public/private coalition had not established a county toll road authority to build the outer loop in order to free up state money to spend elsewhere, he says, "we'd still be waiting for money from the state." Trietsch also credits the Texas Transportation Institute, which has been in a close working relationship with Houston officials for more than 30 years, for helping further the spirit of innovation in the region's transportation function.

In Houston, cooperation between transit and highway officials demonstrated an across-the-board commitment to mobility concerns. And the Harris County delegation to the state legislature, which before the RMP process had been so divided that transportation in Houston was essentially unrepresented at the state level, came together and spoke with a nearly unanimous voice in Austin on transportation matters.

Houston's successful program to improve mobility was a product of creativity, hard work, and some unique features. Some of the lessons it has for other communities include the following:

Having a critical problem helps. A rapid decline in mobility that received national media coverage and was corroborated in independent studies convinced Houstonians that their congestion was the worst in the United States. The Chamber of Commerce's own studies ranked congestion in Houston just behind congestion in Los Angeles.[11] A crisis mentality spurred the community to extraordinary efforts.

So does a simple intergovernmental context. Because of easy Texas annexation laws, the city of Houston accounts for most of the region's population and virtually all of its jobs. A relatively simple intergovernmental context—one major city, a single county, one transit operator, and a state transportation department—allowed the formation of a small Supergroup (see later) that could call the shots effectively.

Options should not be limited to existing funds. Before the RMP process got underway, elected officials and business leaders concentrated on coping within existing funding limits. They did not want to have to talk about tax increases. During the RMP process, Houstonians discovered that a scarcity of resources for transportation was a national problem. If the problem is serious enough and affects the entire community, taxes can be raised and other funding sources—like state transportation departments—can be tapped.

The agencies that are involved must really work together. Just talking is not enough. This is one of the oldest maxims in transportation—and one of the most ignored. Federal requirements for cooperative

planning date back at least to the 1950s. In Houston, as one observer pointed out, the problem was not that these guys don't know each other. They've been talking for years. But each agency had its own agenda, and they all competed. It was the business community that brought them together to really work: the carrot was that all the agencies would get more money. The interagency communication process has been ratcheted up another notch with the establishment of the TranStar Center, which makes working together a daily necessity.

A can-do, business-led campaign can turn the tide. This approach works especially well in communities in which business is respected and government is suspected. For business to take the lead on such a visible public issue—in partnership with the public sector—was normal in Houston. In this instance, the business community validated the importance of the problem and refused to let the predictable doom and gloom views of transportation engineers and bureaucrats dominate the process. It offered a rigorous analytical framework for quantifying the problem and the cost of solving it. And, it established a time frame for the proposed transportation program. Also, business leaders were effective advocates for the needed improvements before the public and the legislature.

A large-scale and long-term effort requires leadership. Many leaders helped bring Houston's regional transportation vision to consensus. Developers and other members of the business community were there at the right time to lend critical support. Bob Lanier, known to some as the high sheriff of transportation, is a student of cities and former developer who receives high marks from virtually everyone for his role in the RMP effort, as chairman of the mobility planning committee, and beyond—as chairman of the state Highway Commission, chairman of METRO, and mayor of Houston. His unusual resume has given transportation in Houston a rare opportunity, compared with most other cities. For that reason, the end of his term as mayor in 1997 is an event that concerns transportation advocates.

There is no single solution to transportation problems. This old chestnut was found to be valid in Houston. The first RMP planners came to the realization that highways alone could not solve Houston's mobility problem. This realization that the transportation system had to be multimodal must have come as quite a shock in the oil capital of the free world. Houston has made a serious investment in transit, and transit will continue to receive one-fifth of the regional transportation investment, even though its share of the travel market is only 1.9 percent (but its share of the downtown commuter market is a more respectable 29 percent). However, the prominence of transit in the plan has also caused problems. According to Roger Hord, many members of the business community are unhappy about the battle over a rail project and they regret the lost opportunities to plan and execute different projects. Houston is clearly divided over the prospects for rail, and the Chamber of Commerce would have taken a different approach to the issue if it had recognized this earlier.

There should be a measurable objective. Houston's goal—to improve congestion to below 0.9 on the TTI roadway congestion index—was clearly stated and eminently measurable. It made reassessing the program at midterm and making midcourse corrections relatively easy. With such obvious success at implementing many of the objectives, it would have been easy for all involved to become complacent—had they not had a specific goal in mind.

Flexibility is important. Through many changes in Houston's transportation plans—none greater than the defeat of the rail plan, which Warren Bold, in supreme understatement, calls a midcourse adjustment—the consensus held. The strength of the coalition made major shifts possible.

Keeping the process simple helps. Houston's ad hoc effort—led by a Supergroup that met in Bob Lanier's mansion in River Oaks and resembled an old boy's club much more than a comprehensive, participatory planning process—seemed more effective than traditional public planning efforts intended to accomplish the same kind of goals. One reason why traditional planning efforts work less easily is that they tend to be loaded down with many and often competing transportation agency agendas.

And speaking of the Supergroup. If the transportation planning and implementation process in Houston was to work, all the agencies with a hand in the solution would have to be brought to the table. The Supergroup, an ad hoc coalition of the community's highest elected and appointed officials and its prominent civic leaders, was the instrument that evolved. The Supergroup is unmatched in the United States as a tool for achieving consensus on mobility issues. It includes the mayor of Houston, the chief elected official of Harris County, the chair of METRO, the chair of the state Highway Commission, and the chair of the Greater Houston Chamber of Commerce Committee for Regional Mobility. During the crisis period, this group met every six to eight weeks, and dealt only with transportation. The active involvement of these top officials facilitated the interagency cooperation that was so important to the success of Houston's mobility program.

Notes

1. Robert T. Dunphy, "Houston Takes a Businesslike Approach to Regaining Mobility," *Urban Land,* March 1993, pp. 31–34.

2. Donald L. Williams and Sally M. Dwyer, "Houston Real Estate Battles Back," *Urban Land,* May 1992, pp. 25–40.

3. Committee for Regional Mobility, *Regional Mobility Plan for the Houston Area: 1989* (Houston: Greater Houston Partnership, 1989); and Transportation Committee, *Regional Mobility Plan for the Houston Area: 1982* (Houston: Houston Chamber of Commerce, 1982).

4. Kathy Whitmire, remarks at a conference (ULI spring meeting, Houston, May 1992).

5. The 1990 Census calculates a somewhat different commuting share because it uses a different definition of the Houston region.

6. Employer Trip Reduction Task Force, *Commuting Experience, Alternatives, and Preferences* (Houston: Central Houston, Inc., 1993), p. 3.

7. *Twelve Tools for Improving Mobility and Managing Congestion* (Washington, D.C.: ULI–the Urban Land Institute, 1991) contains some discussion of these concepts.

8. Houston-Galveston Area Council, *Access 2010: 1994 Update,* executive summary (Houston: Houston-Galveston Area Council, December 1994), pp. 4–5.

9. Bruce Schoenfeld, "Core Belief," *Profiles* (Continental Airlines), April 1996, pp. 27–28.

10. Ibid., p. 29.

11. This experience was confirmed in a series of seminars on suburban mobility that was sponsored by the U.S. Urban Mass Transportation Administration. Participants in the seminars knew that transportation was a problem, but they were unaware of its magnitude or complexity. See Frederick W. Ducca, "UMTA's Suburban Mobility Seminars: The Education Process," *ITE Journal,* February 1992, pp. 36–39.

Chapter 11

Findings and Outlook

In the end, all solutions are temporary.

— Garrison Keillor, extolling the virtues of duct tape on *A Prairie Home Companion*

This study offers a message of hope for people engaged in the development and traffic dilemma. Success happens. The common view that, like the weather, there is basically nothing that can be done about traffic is a misperception.

But this success is not an off-the-shelf commodity. Each solution must be handcrafted. The successes in the case studies in this book are all very different from each other. Each community must find a solution that works. Boosters of a particular technique or region like to proclaim their way as the one way. The urban regions studied here, however, have taken different paths to resolving traffic and development challenges, ways that work for them.

Just as it would be inappropriate to pursue a Houston style mobility program in Portland, it would be equally goofy to impose a Portland style urban growth boundary and highway limits in Houston. Approaches that work in one region may not work elsewhere because of sharp differences in attitudes and cultures. Rule number one for a community dealing with traffic and development problems is: know thyself or know thy residents—-their perceptions on what the problems are, their taste for different potential approaches, and their willingness to provide for growth.

What Is Success?

The case studies in this book were selected because they were deemed to be good examples of how to manage growth and transportation at a regional level. The determination of what are good examples was admittedly highly subjective, based in part on objective statistical measures and partly on the degree to which the region's planning and transportation policies were in harmony. It was decided to limit the

field to communities that have made a difference at the regional level, because growth and development decisions have regional transportation impacts. Many excellent examples of cities, counties, or corridors that have reconciled traffic and development within a messier region therefore were not considered. Most successes in this complex endeavor, in fact, are likely to be found in smaller areas than metropolitan regions. These successes should be recognized, even if they fall short of constituting regional solutions.[1] The sponsors of this study would be happy to hear of undiscovered successes.

Objective Measures: Congestion and Mobility

A reduction in congestion would be a simple measure of success. According to data compiled by the Texas Transportation Institute (TTI), only three urban areas achieved a reduction in congestion between 1981 and 1992: Houston and Phoenix through transportation investments, and Detroit through a deteriorating economy. Congestion grew during this period in most of the case study regions, but at the same time, transportation investments were being made and planning was being undertaken to address congestion.

Congestion and mobility are the yin and yang of transportation performance. In fact, neither congestion nor mobility is an entirely objective concept. And both are notoriously difficult to measure.

The TTI congestion index, which expresses the ratio between traffic volumes and highway capacity, is admittedly simplistic. The public's perception of congestion may be as important as technical measurements of it, and the two sometimes do not agree. If the public thinks that there is a traffic problem, there is a traffic problem. Congestion is an issue in St. Louis, but the data say that it should not be. In

Phoenix, congestion has been reduced slightly and the residents seem to think that it is no longer an issue.

The TTI index, which is the only recognized standard for congestion, fails to deal with modes of travel other than cars, which is a major weakness for places like Manhattan or downtown San Francisco. Another problem with the TTI index is that its assessment of highway capacity does not take into account improvements, such as intelligent transportation systems, that help traffic move better. TTI is studying how to improve the index.

Mobility is a sometimes difficult concept, especially as a goal of public policy. From the traveler's perspective, mobility may not mean moving fast. Take Susan Handy, a professor of community and regional planning at the University of Texas: she describes her shock at moving from Berkeley, where motion was slow but the array of nearby choices was rich, to Austin, where one can drive at high speeds but cannot easily find a latte bar. It is richness of choice in culture, entertainment, and shopping that makes for a great place (like Berkeley, Seattle, or Manhattan), even if the speed of automobile travel within it is snaillike. Such places require a sensitive balance of land uses, travel choices, and travel speeds that is not easily quantified. As a minimum, any harmful imacts of transportation investments on such urban settings should be avoided; at best, transportation investments that improve great urban places should be made.[2]

A Harmonic Convergence between Planning and Transportation Policy

The other measure that was used to select case studies was the degree of consistency between planning and transportation policies. Achieving such consistency is extraordinarily difficult. Different levels of government are involved, different agencies, and different professional disciplines (typically engineers versus planners).

But when this harmony is missing, the usual result is gridlock in the planning office. Development decisions are held up because the transportation facilities are inadequate to support development. Transportation decisions are held up because transportation improvements might encourage development in the wrong places. Public policy gridlock can be even more maddening to the participants and to the public than the gridlock experienced on the roads. By its nature, planning policy does not lend itself to quantitative evaluation. However, it is generally observed that the regions that are included as case studies in this book have achieved such harmony, even if some of the policies themselves have chinks.

Results in the Case Study Regions

Each region included as a case study has its own particular context, but for the purpose of drawing usable lessons from their various experiences the seven regions can be sorted into four primary but not mutually exclusive categories: livable community (Toronto, Portland), resort-oriented community (San Diego, Phoenix), boom town (Atlanta, Houston), and older industrial area (St. Louis).

Livable Communities. For livable communities like Portland and Toronto, the quality of urban life and the quality of the environment are important values. Different governmental jurisdictions appear to be willing to cooperate in order to improve the larger region. Many of the approaches taken to transportation and growth issues in these two case study regions are similar.

In Portland, the initial focus was on the revitalization of the city of Portland. That battle has been won in large part, and the focus has shifted to suburban policies that can help create a livable region. Beginning in the 1970s, a series of transportation initiatives that supported downtown's revitalization were undertaken: the elimination of downtown freeways from roadbuilding plans, the removal of a road along the riverfront, the imposition of a lid on downtown parking, the development of a successful downtown bus mall, the opening of a light-rail line, and bus service improvements. The establishment of an urban growth boundary also supported the orientation to downtown, while it protected land resources and helped stem some of the worst excesses of suburban sprawl development.

Downtown Portland has become a thriving center for business, shopping, and dining. And transit has maintained its traditionally high share of downtown travel. A significant amount of development along the new light-rail line has occurred, although how well some of the projects are related to transit is difficult to tell. Recent apartment development activity along the existing line and the future west side line is a promising sign in this regard.

The development of the suburbs is a different type of problem that will require the same degree of creativity. The urban growth boundary is an extraordinary tool, but one that needs to be used more in the service of higher-density infill development rather than in the service of sprawl development within the boundary. The growth boundary has protected local homebuilders from the competition of national firms, which consider the available subdivision sites too small for their scale of operations. But many of these local builders have concentrated on large houses for traditional families.

Portland's aggressive policies to reduce driving have not prevented a growth in highway travel that has significantly outpaced the increase in transit use. Traffic congestion appears to be growing rapidly. Far from falling into despair at such failures, however, Portland's leaders, it seems, are intensifying their search for new ways of meeting transportation goals. Politics and policies, says the regional government's assistant transportation director Keith Lawton, are out ahead of the technical process; policies to reduce vehicle-miles traveled have been adopted despite the fact that no good mathematical models for testing their assumptions are available.[3] Policy makers may be criticized by transportation modelers, but they feel that they have won the battle of public opinion. Residents support the growth boundary and other attempts to manage growth, and they love MAX, the light-rail line, and a large majority is willing to support it with taxes. Business seems also strongly supportive of the direction that planning is taking in the Portland region, and the region's multijurisdictional governance structure will serve it well in dealing with future issues and problems.

Toronto, with four decades of history in livable community management, stands in sharp contrast to most U.S. regions. The region boasts an urbane community, a well-used public transportation system that offers excellent mobility, an inner city that attracts upper-income residents, and a system of governance that emphasizes the concerns of the broad regional community.

While the Toronto region represents a true success story, the nature of its growth has shifted in recent years. The new parts of Toronto are increasingly distant and different from the central region. Newly developed areas are less urbane, less densely settled, and more auto-oriented. The problem is widely recognized, and leaders are aggressively trying to deal with it. Toronto's long tradition of good government and good planning suggests that the challenge will be met.

Resort-Oriented Communities. With their favorable climates and natural attractions, the San Diego and Phoenix regions are oriented to vacationers and retirees, while they both also have attracted much business development and gained an extraordinary number of permanent residents. Population growth and economic development have threatened the very qualities that make these regions such attractive destinations. These regions are also similar in terms of the high value they place on cost-effective government.

San Diego's history of strong planning is almost as old as Portland's. Within the city, regulations to limit the worst excesses of sprawl and encourage growth in existing urban areas have been accompa-

nied by some striking successes in downtown revitalization. Although the region lacks a true regional government, it has been able to mobilize for transportation improvements.

But, unlike Portland, San Diego has not made a clear choice in favor of transit over roads. Transportation spending on transit is higher than might be expected, given transit's market share. As a whole, however, the region is dominated by the automobile. Despite California's strong antitax movement, San Diego's voters approved an added sales tax to finance the new roads and other infrastructure required for growth. When the special tax expires, the region will be challenged to find additional transportation financing.

In terms of growth over the past two decades, Phoenix is in the same league as San Diego, but it has not enjoyed the luxury of a freeway system provided by the state. Transportation needs are intensified by the tendency of subdivisions to pop up all over the desert, sometimes, it seems, overnight. Residents' concerns about unlimited sprawl conflict with their western ethic of private property rights and freedom from regulation, but these concerns were powerful enough to lead Phoenix to adopt a concept of more managed growth, which it termed a system of urban villages.

While the urban village program lacks the types of planning controls and regional governance found in Toronto and Portland, it represents an important effort to create some order out of the chaos of uncontrolled rapid growth. The vision predates some of the neotraditional planning concepts that took hold in California in the 1990s. With the aid of a dedicated funding source—another sales tax—Phoenix also embarked on a massive freeway building program in the 1980s, which, although not fully carried out, was another important initiative for dealing with growth needs in a regional context.

Boom Towns. Atlanta and Houston are more typical examples of fast-growing southern and western regions, in that their growth tends to take place within a boom-and-bust cycle. These two boom towns are unabashed self-boosters. They are progrowth and probusiness, and they tend to lack the stomach for managing growth in the fashion of livable communities or resort-oriented communities.

Atlanta fairly bursts at the seams with new people, new businesses, and new enthusiasm. It contains some of North America's foremost edge cities. As in Toronto, the central region—which has a long tradition of regional cooperation—must cope with growth outside its purview. Business is moving without constraint to ever more distant centers and population growth is occurring beyond them.

During the 1980s, Atlanta developed an effective public transit system and opened a successful subway. However, the true region continues to move well beyond transit's service area. At the same time, the region's aging freeway system was completely rebuilt and widened under the auspices of the state DOT. The roadbuilding effort was extraordinarily effective—for a while. Transportation needs have expanded with the growing scale of Atlanta, and the region faces a greater challenge than it has ever faced in this regard.

In Houston, the lack of zoning seems to be of little concern and initiatives to enact it have been regularly defeated. In the early 1980s, with a strong push from the business community, Houston met the challenge of congestion with a program to improve mobility rather than with a program to control development. The investment program involved a truly Texas-size bill, making Houston among the case study regions the one most dedicated to a supply-side solution.

Contrary to popular perceptions, the investment program earmarked a considerable amount of spending for transit, and transit has captured a growing share of travel. In certain transportation corridors, the capacity for building more roads may be limited. But Houston's leading-edge, high-tech traffic management system represents a good hedge for a future of limited new construction.

Older Industrial Communities. St. Louis is the case study representing cities without recent healthy growth, especially older Rustbelt industrial communities. Its latest transportation success story is the MetroLink light-rail line, which although newer already carries more passengers than Portland's MAX. The regional transit agency plans to extend the MetroLink system around the region, using it as a tool for directing growth, one of the few such tools available in this conservative region.

St. Louis may not have grown in population in recent years, but it has gained a significant number of jobs and highway travel has grown accordingly. Many successes have been achieved in the arena of rebuilding the central city, although out-migration is continuing. The transit agency has become a credible institution in the region, which means that it may be able politically to take on additional responsibilities, including the promotion of rebuilding around transit stations in key neighborhood in the central city, where transit works best.

All Success Is Temporary

Success may not last. In the post–World War II period, Houston enjoyed a high level of mobility. Then the region was swamped with growth. Now, it is again successfully regaining mobility. In each success are the seeds of defeat, and constant vigilance is required. Toronto adopted planning policies in the 1950s and 1960s that made it a model region. Today, it faces some serious challenges to its quality of life as the formerly successful planning policies no longer seem relevant.

The Transferability of Case Study Models

The successful transfer of lessons requires an appreciation of local circumstances. The case studies in this book represent markets that are generally in the middle range of large metropolitan areas in terms of population, which ranges from just under 2 million to about 4 million. Small regions tend to borrow from large cities and regions, in the belief that bigger models are a better source of ideas.

While many of tools—such as transit-oriented development, HOV lanes, and rail transit—that are described in these case studies may not be appropriate for small cities, they should find the concept of compatible planning and transportation policy useful. In small towns and urban regions, the quality of life may be even more endangered through poorly executed transportation projects and poor planning than it is in large regions.

Can other large urban areas learn from each other? Certainly, good planning requires an understanding of state-of-the-art practices in other regions. The regions examined in this book are less complex than the largest U.S. urban regions. The Phoenix, San Diego, and Houston regions consist primarily of one county with one large city and a number of smaller cities. Only St. Louis among the case studies has to contend with two different states. In more politically and culturally diverse regions, the process of reaching a regional consensus is more difficult than it was for many of the regions considered in this book. But that should not diminish the accomplishments of these regions.

Rick Cole, director of the Local Government Association of Southern California, says that people in Los Angeles are interested in what Portland is doing, but that they "wonder if the fog and the Birkenstocks make them somewhat different." In Los Angeles, in fact, an effort to identify best practices around the country is underway as part of a livable communities initiative.

Regional Approaches, Not Regional Government

A regional approach to development and transportation issues does not necessarily require a regional government. The regional governments in Portland

and Toronto, in fact, stand virtually alone. Therein is the problem with proposing regional government as a solution to the development and transportation dilemma: there does not seem to be any political support for regional government.

During the 1960s, federal legislation established regional agencies to carry out metropolitan transportation planning and the U.S. Department of Housing and Urban Development helped fund many metropolitan planning agencies. When federal funding receded, so did much of the interest in these organizations, especially for those dealing with comprehensive planning. Concern over regional issues has lessened also with the population (and power) shift from city to suburb. And changing work and shopping patterns have lessened the interdependence between cities and suburbs, and further lessened support for regional governance. Even in Canada, competition for tax base among cities in regions is on the rise.

Fortunately, regional approaches do not require regional government. In St. Louis, the council of governments initiated the concept of a light-rail line, passed it off to operating agencies, and worked behind the scenes to get it built—a classic example of achieving results without authority. In part because of this success, the council of governments appears to have gained credibility among state transportation officials. In San Diego, the association of governments was given authority by local governments to manage growth policy and to keep the purse for transportation revenues. This is a rare example of a voluntary agency gaining spending authority.[4]

Houston provides an example of major agencies working together on a voluntary basis. Putting together an unofficial intergovernmental coalition seems to have worked for that region.

On the other hand, Portland's Metro, a true regional government, shows that having legal authority is sometimes not enough. Success in carrying out a regional vision requires reaching out to local governments, citizens, the business community, and state agencies.

Federal and State Transportation Agencies

When regions attempt to achieve consistent transportation and development strategies, they often find the effort complicated by the operating procedures and transportation policies of state and federal transportation agencies. State and federal funding combined finances 75 percent of U.S. roads and 65 percent of U.S. transit spending. These agencies represent an essential source of revenues, a technical resource, and a set of regulatory standards.

On the land development side of the traffic/land use dilemma, so far the federal government and most states have stayed out of land use planning, which is a jealously guarded local prerogative. On the federal level, the political winds seem to be blowing toward deregulation and devolution of authority to states and localities. State and provincial policy on regulating land use seems to be blowing in similar directions, but not universally. Canadian provinces exercise greater influence on regional housing policies and planning consistency than do U.S. states in general.

Some states are becoming more active in directing growth and infrastructure policies, usually focusing on planning requirements rather than outcomes. The state of Oregon has probably been the most aggressive. It requires localities to set growth boundaries and follow certain transportation and parking guidelines. But legislative support for such measures has not always been strong. Georgia and Arizona not only provided funding but also planning capacity and their status as owners of the main roads, to help Atlanta and Phoenix, respectively, upgrade their transportation systems.

Few states will continue to play the role of disinterested party in regional transportation plans. The future of the federal role will be determined by the reauthorization of ISTEA in 1997.

The Environmental Imperative

Environmental concerns have always been an important element of transportation and development policies and practices. The 1990 federal Clean Air Act Amendments, for example, require regional transportation planning to achieve air quality standards, and the amendments even make some inroads into land use regulation. Polls show broad public support for environmental quality. The easy fix is the technological one. In fact, the average car today produces 90 percent less pollutants than it did 30 years ago. Transportation projects that reduce congestion are also beneficial because they result in less pollution—although some critics think that making driving easier simply perpetuates an auto-dominant lifestyle. Atlanta's freeway program helped improve air quality. Transit improvements are thought to help the environment by getting people out of cars—although some critics point out that cars are more energy-efficient per capita than buses, considering the increased fuel efficiency of automobiles and the low average levels of bus loading. Portland's transit program helped that region improve air quality.

Similarly, certain sprawl development patterns are seen as environmentally harmful while higher-density, compact development that offers a choice of travel modes is promoted as environmentally sensitive. As with transportation practices, the real need is to get behind the claims, make objective evaluations of environmental impacts, and find workable solutions. Clearly, any solution must meet high standards of environmental quality, in keeping with popular proenvironment sentiments.

Keys to Success

The success stories in this book should be studied individually for possibly transferable ideas. Together they offer some underlying themes for people involved in development and transportation issues, as follows.

A Consistent Vision

Major transportation decisions are difficult in and of themselves. They involve large amounts of money and transportation facilities use large amounts of land and can disrupt neighborhoods. Transportation decisions that are inconsistent with regional development patterns can create planning gridlock, where nothing can be accomplished. Since metropolitan transportation planning was officially sanctioned by the federal government, transportation plans have had to be consistent with community goals.

Now that ISTEA has applied a more specific standard of consistency and now that communities (and regions) engage in more visible disagreements on growth goals, it is much more difficult to advance major transportation facilities. Proposing a growth-inducing transportation project in a corridor that is expected to remain undeveloped is a recipe for disaster. State departments of transportation prefer to stay out of such battles, which is one of the reasons that they support metropolitan planning organizations. A transit plan was advanced in Seattle before the region had determined some growth policies as required by state legislation. The plan was criticized for this shortcoming. In Phoenix, the urban village plan is being rethought, which creates a kind of vacuum for planning transportation projects.

Knowing the Territory

Transportation has strong political overtones, but transportation practitioners, who often are engineers, are generally uncomfortable with the unquantifiable aspects of political maneuvering. Planners need to make sure that their plans represent ideas with which the community is comfortable, which is one of the reasons that the federal government imposes requirements for citizen participation. Transportation engineers certainly do not wish to deal with irate citizens and officials. San Diego represents a good example of a technical/political planning process.

The plan began with the mathematical models that provided technical authority and then was carefully gone through to find a mix of projects that could be implemented.

Supporting a transportation project on the basis that it is the right thing to do is an exercise in frustration. That is a great weakness of peak hour pricing, widely seen by economists as the right thing to do. Pricing options have considerable technical merit and a single drawback: people generally hate them.

Knowing the territory means knowing how bad the transportation problem is, not in terms of objective measurements but in the eye of the public. Measurements can help define the problem and possibly explain it. But solutions require a willing public. An outsider finds it hard to see that St. Louis has a traffic problem, but the region's citizens believe traffic is bad enough to justify paying for improvements. (Admittedly, St. Louis has pockets of severe congestion, but they are fairly limited.) The strongest support for transportation improvements in Houston came from the widely held belief that Houston suffered the worst congestion in the United States, and something had to be done to fix it.

This fixation on problems is anathema to planning, but it is a reality of the American psyche. The flip side of the focus on problems is that the public loses attention when things get better, a current concern in Houston and San Diego.

Reliable Funding

All the many techniques used by the seven regions studied to improve planning and transportation needed funding. San Diegans voted a local sales tax to provide resources. Houstonians put together a package of tolls, local revenue sources, and gas tax increases. In Toronto, lack of funding for major transit improvements has been a sore spot for a number of years. Large funding increases in Phoenix and Atlanta made a significant difference in transportation projects, but they have since been cut back.

Innovative financing—which means, according to some critics, "I don't have any money, do you?" —has captured the attention of transportation planners in recent years. This search for alternatives indicates the declining significance of traditional finance, especially the gasoline tax and other charges on users. These traditional means of financing have the advantages of simplicity and general equity. People who drive more pay more. User charges are collected at the pump (usually) or at registration time, generally efficiently. States with lower roadbuilding costs can have a lower gas tax, making it even more attuned to the costs of travel. Transit fees accomplish

Figure 11-1
New Light-Rail Systems, 1995

| | Line-Miles | Weekday Riders | | Capital Cost[1] | | |
		Daily Average	Per Line-Mile	Total	Per Line-Mile	Per Weekday Roundtrip[2]
Calgary	18.2	110,000	6,044	$560,000,000	$30,800,000	$10,182
Edmonton	7.6	36,000	4,737	229,000,000	30,100,000	12,722
Denver	5.3	14,000	2,642	109,000,000	20,600,000	15,571
St. Louis	17.0	40,000	2,353	361,000,000	21,200,000	18,050
San Diego	34.5	44,000	1,284	388,000,000	11,200,000	17,517
Sacramento	18.3	23,000	1,257	225,000,000	12,300,000	19,565
Portland	15.1	24,000	1,649	284,000,000	18,800,000	22,811
Baltimore	22.5	20,000	889	449,000,000	22,200,000	44,900
Los Angeles (Blue Line)	22.0	38,000	1,727	963,000,000	43,800,000	50,684
Buffalo	6.4	29,000	4,531	751,000,000	117,300,000	51,793
Santa Clara County	21.0	20,000	952	641,000,000	30,500,000	64,100

[1]As of 1994.
[2]Assumes average rider makes two trips per weekday.
Source: BRW Inc.

the same thing, although the share of total costs collected from the user is much lower.

It matters little to the user whether the gas tax is federal, state, or local. It matters to the state depending on whether or not it is a donor state (one that gets back fewer federal transportation dollars than the federal government collects in the state). Trucks are very important in highway financing, accounting for around 40 percent of highway revenues.

A gas tax is very different from congestion pricing, in that the user pays the same cost per mile whether driving on congested or traffic-free roads (not counting the gas that is wasted by driving in stop-and-go conditions). Several of the regional transportation investment programs among the case study cities did not rely on the gas tax as the primary financing source. San Diego and Phoenix used sales taxes, a scheme under which road hogs and people without cars pay the same amount (assuming similar consumption habits). The gas tax seems to suffer from a political liability, not a funding one. In that people view it as a tax rather than as a user fee, legislators are often loath to increase it.

An ideal financing system would charge users for the incremental costs they impose on the transportation system—driving on crowded routes during peak hours would cost more than driving on free flowing routes during off-peak hours. The tolls on a new four-lane section of SR91 in Orange County, California, vary from $2.50 to $0.25, depending on the time of day. This is a demonstration program. Such

pricing could affect demand, by making peak hour travelers reconsider the time and the route of the trip, or even the trip itself.

At the other extreme are sales taxes and development fees, which charge the same regardless of how much or little a person drives. Development fees have the further limitation that they represent an upfront contribution for a service that will be used over a period of possibly 20 years, a little like paying upfront for 20 years of groceries.

Getting Serious about Transit

Each of the communities studied here has paid significant attention to transit improvements. For three of them—Toronto, Portland, and San Diego—transit is a central focus of the regional development policy. St. Louis has used its light-rail corridor to restructure its bus system as well. Atlanta has the only newly constructed heavy-rail system. MARTA is a project intended to promote mobility and the downtown. Most of the case study cities have emphasized downtown revitalization, which strengthens the prime transit market. Houston and Phoenix, two cities that are not generally known for their transit systems, have made significant bus improvements that have produced the best relative gains in transit ridership among the seven regions, and some of the highest gains among all regions.

Getting serious about transit requires a realistic assessment of the potential transit market and the regional development plan. Focusing growth into transit corridors certainly helps build the transit market. A transit alternative is essential not only for its political correctness, but also for its contribution to the health of the central area. (It also demonstrates that other options were considered along with highway projects.) Less sophisticated transit supporters develop a fixation on facilities and usually jump on the light-rail bandwagon. Experience indicates that light rail can be an expensive means of serving transit riders. Some lines cost over $100,000 per daily rider served. Moreover, beyond the initial expense, ongoing subsidies are required. Some light-rail advocates claim that such systems are cheaper than the alternative in urbanized corridors. That may be true, but the claim should be checked out for each corridor. As James Mills of San Diego says, "the issue is not a rail system, but what mode works best for each corridor."

Taming the Car

Taming the car means squeezing as much efficiency as is possible out of the current transportation system before considering the expansion of roads. It may be possible to introduce greater controls on parking, as was done in Portland, or impose higher road-use fees during peak hours, as has been tried only in demonstration programs. One interesting twist on congestion pricing is being considered in San Diego and Houston, an HOV buy-in. This allows solo drivers to pay to use HOV lanes and will work as long as traffic speeds remain high.

HOV lanes offer an in-between option for increasing the efficiency of a highway corridor in which there is significant demand. Some environmentalists oppose them, saying that they are often just a stalking horse for new highway facilities and that their occupancy rules will not be enforced.

There is a growing sense that the building era for highway improvements is over. As with many of the generalizations in transportation, this one should not be accepted uncritically. Most areas have growing, undeveloped suburbs where highway improvements are not only possible, but also necessary. Furthermore, there will continue to be strong political support— if not financial support—for highway improvements.

The Need to Follow Up

Travel behaviors change, economies change, and new issues come to the forefront. Major plans need plenty of follow-up and, ideally, a feedback mechanism that permits them to be regularly adjusted. In Houston, for example, a planned midcourse review of the regional mobility plan to check up on the progress of implementation and consider new issues made a significant change: to drop plans for a rail system and reorganize around a bus plan. Canadian planning does a better job than U.S. planning at follow-up, with regularly scheduled plan updates. In Toronto, the multicentered plan concept that was introduced in 1980 was expanded in 1994 to reflect major changes in transit projects.

Portland is a good example of a region that maintains a strong tradition of planning and follows up over a long period of time. For many elements of transportation and land use planning, it is important to ask: Does it sell? If transit ridership is too low, freeway traffic too high, or congestion pricing unsuccessful, the plans can be adjusted to the market. The market evaluation is even more important when promoting development concepts that are new to the market. In the end, land use solutions must find buyers. If transit-oriented communities do not sell, the design or the concept must be adjusted until it works. In the end, the choice is up to the user.

Muddling Through as a Fallback Position

What is a region to do when it realizes that it will never make the exalted "A" list of regions that have successfully met the challenges of transportation and development? In an excellent review of the transportation/land use connection, Terry Moore and Paul Thornes recognize this reality. Regions that are not able to muster support for an impressive package of improvements can, they suggest, squeeze more capacity out of the existing transportation system, improve regional coordination on land use and transportation issues, and use more effective pricing policies for infrastructure.[5] Anthony Downs reviews a long list of improvements and concludes that the ones that would be most effective—road pricing and parking surcharges—are not likely to be adopted because of political opposition.[6]

A nonaggressive approach to traffic growth might include equal parts of improved efficiency, added capacity, and increased congestion. A one-third share of the growth could be accommodated simply through more congestion, while a program of capacity improvements and more effective traffic operations would handle a two-thirds share. The answer for transit may be similar: some improved services, some higher prices, some crowding, and some longer wait times.

Notes

1. The successful application of various techniques, state-of-the-art practices, and new ideas is the focus of much of the professional transportation literature. See *Twelve Tools for Improving Mobility and Managing Conges-*

tion (Washington, D.C.: ULI–the Urban Land Institute, 1991) and *ITE Toolbox for Alleviating Traffic Congestion* (Washington, D.C.: Institute of Traffic Engineers, forthcoming 1997) for publications intended for a nontechnical audience.

2. Susan Handy, "Highway Blues: Nothing a Little Accessibility Can't Cure," *Access* (University of California Transportation Center, Berkeley), fall 1994, p. 4.

3. Keith Lawton, interview by author, September 25, 1996.

4. Another example is the Metropolitan Transportation Commission in the San Francisco Bay Area, which has long had authority from the state to distribute transit funds.

5. Terry Moore and Paul Thornes, *The Transportation/Land Use Connection*, Planning Advisory Service Report no. 448/449 (Chicago: American Planning Association, 1994).

6. Anthony Downs, *Stuck in Traffic: Coping with Peak Hour Traffic Congestion* (Washington, D.C.: The Brookings Institution; and Cambridge, Massachusetts: Lincoln Institute of Land Policy, 1992), p. 154.